Internet of Things with the Arduino Yún

Projects to help you build a world of smarter things

Marco Schwartz

BIRMINGHAM - MUMBAI

Internet of Things with the Arduino Yún

First Published: May 2014

Production reference: 1140514

Published by Packt Publishing Ltd.
Livery Place
35 Livery Street
Birmingham B3 2PB, UK.

ISBN 978-1-78328-800-7

www.packtpub.com

Cover Image by Shweta Suresh Karkera (shwetaimages@gmail.com)

Credits

Author
Marco Schwartz

Reviewers
Fiore Basile
Charalampos Doukas
Francis Perea

Commissioning Editor
Anthony Albuquerque

Acquisition Editor
Harsha Bharwani

Content Development Editor
Poonam Jain

Technical Editors
Manan Badani
Shashank Desai
Shali Sasidharan

Copy Editors
Tanvi Gaitonde
Dipti Kapadia

Project Coordinator
Melita Lobo

Proofreaders
Maria Gould
Ameesha Green

Indexer
Hemangini Bari

Production Coordinator
Alwin Roy

Cover Work
Alwin Roy

About the Author

Marco Schwartz is an electrical engineer, entrepreneur, and blogger. He has a master's degree in Electrical Engineering and Computer Science from Supélec in France, and a master's degree in Micro Engineering from the EPFL in Switzerland.

Marco has more than five years of experience in the domain of electrical engineering. His interests gravitate around electronics, home automation, the Arduino and Raspberry Pi platforms, open source hardware projects, and 3D printing.

About the Reviewers

Fiore Basile is a programmer, system administrator, creative, entrepreneur, and maker. Since 1996, he has served as a project manager, consultant, and technology officer in industrial and research projects of varied sizes across Italy and Europe. He has worked in the fields of cultural heritage, e-health, digital preservation, multimodal interfaces, and web and mobile publishing. During his career, he has also started two IT start-ups, held workshops at international conferences and events, and has been interviewed by national and international press. His work experience allowed him to build a broad expertise in systems, web and mobile software development, open source and open hardware, embedded programming, and electronics. He's currently conducting research on wearable technologies, affective computing, and smart connected devices. He also works as the coordinator of FabLab Cascina, a digital fabrication laboratory in the middle of Tuscany.

Charalampos Doukas is a researcher and an IoT maker. He started playing with sensors and Arduinos in 2008 when trying to capture and transmit vital signs. He is passionate about combining different hardware systems with software and services using the Internet. He helps in spreading knowledge about open source software and hardware by organizing sessions in workshops and conferences.

He has built many projects around home monitoring and automation. He contributes hardware nodes for Node-RED and has also authored the book, *Building Internet of Things with the Arduino, CreateSpace*.

When Charalampos is not playing with sensors and actuators, he manages European research projects at CREATE-NET in Trento, Italy.

Francis Perea is a professional education professor at Consejería de Educación Junta de Andalucía in Spain with more than 14 years of experience.

He specializes in system administration, web development, and content management systems. In his spare time, he works as a freelancer and collaborates, among others, with ñ multimedia, a small design studio in Córdoba, working as a system administrator and main web developer.

He also collaborated as a technical reviewer on the book, *SketchUp 2014 for Architectural Visualization, Thomas Bleicher and Robin de Jongh, Packt Publishing*.

When he is not sitting in front of a computer or tinkering in his workshop, he can be found running or riding his bike through the tracks and hills in Axarquía County, where he lives.

I would like to thank my wife, Salomé, and our three kids, Paula, Álvaro, and Javi, for all the support they gave me, even when we were all busy. There are no words that would be enough to express my gratitude.

I would also like to thank my colleagues in ñ multimedia and my patient students. The need to be at the level you demand is what keeps me going forward.

www.PacktPub.com

Support files, eBooks, discount offers, and more

You might want to visit www.PacktPub.com for support files and downloads related to your book.

Did you know that Packt offers eBook versions of every book published, with PDF and ePub files available? You can upgrade to the eBook version at www.PacktPub.com and as a print book customer, you are entitled to a discount on the eBook copy. Get in touch with us at service@packtpub.com for more details.

At www.PacktPub.com, you can also read a collection of free technical articles, sign up for a range of free newsletters and receive exclusive discounts and offers on Packt books and eBooks.

http://PacktLib.PacktPub.com

Do you need instant solutions to your IT questions? PacktLib is Packt's online digital book library. Here, you can access, read and search across Packt's entire library of books.

Why subscribe?

- Fully searchable across every book published by Packt
- Copy and paste, print and bookmark content
- On demand and accessible via web browser

Free access for Packt account holders

If you have an account with Packt at www.PacktPub.com, you can use this to access PacktLib today and view nine entirely free books. Simply use your login credentials for immediate access.

Table of Contents

Preface

The Internet of Things (IoT) is a growing topic in the tech world, and more and more hardware projects that are funded using crowd-funding campaigns include some connected objects. Such objects can be smart watches that connect to the Web, the weather station, cameras, energy monitoring devices, and even robots. Many industry giants such as Google and Samsung are also entering the market with connected objects and wearable devices.

On the other hand, millions of people around the world use the Arduino platform to create hardware projects. Because Arduino is so easy to use, it allows not only hobbyists, but also artists and people without a tech background to create amazing hardware projects. The platform is always evolving with new solutions that allow people to create more and more complex DIY projects.

One of the latest boards from Arduino—the Arduino Yún—mixes these two worlds harmoniously. This board's release was appreciated by hobbyists around the world who wanted to develop connected objects. Indeed, developing applications for IoT has always been quite complex and requires a lot of expertise in both hardware and web applications development. However, we are going to see why using the Arduino Yún can make the process much easier.

The Arduino Yún is the same size as the Arduino Uno, which is the most common Arduino board. However, the difference is that it features a small Linux machine that runs on a separate processor as well as an onboard Wi-Fi chip so you can connect the board to your local Wi-Fi network.

The clever thing they did with the Arduino Yún board is create a Bridge library that allows you to call functions of the Linux machine from the usual Arduino microcontroller that is also present on the board. This way, you can use the powerful features of the Linux machine by programming in the same way as you would on the Arduino Uno board. You can, for example, write whole programs in high-level languages such as Python, and call them from an Arduino sketch.

The fact that the board also has onboard Wi-Fi changes everything. The board was developed in close collaboration with the Temboo web service, which provides many libraries to interface the board with other web services such as Google Docs, Gmail, and Dropbox.

For all these reasons, using the Arduino Yún will allow you to build connected applications without requiring you to be an expert in the field. Using the power of the embedded Linux machine, the Wi-Fi connection, and the Temboo libraries, you will be able to easily create your own IoT devices. To show you what exactly the Arduino Yún can do, I have built four exciting projects using this board, and you too will be able to build these projects after reading this book.

What this book covers

Chapter 1, Building a Weather Station Connected to the Cloud, introduces you to the Internet of Things features of the Arduino Yún. In this chapter, we are going to build a weather measurement station (which measures the temperature, humidity, and light levels) that sends data to the Web. The project will send data to a Google Docs spreadsheet via Temboo and log the results in the spreadsheet where they can be displayed graphically. The nice thing about this project is that this data can then be accessed from anywhere in the world just by logging into your Google account and going to the spreadsheet.

Chapter 2, Creating a Remote Energy Monitoring and Control Device, focuses on energy management by creating a project to switch a device on and off (like a lamp), measuring its energy consumption and storing this data to the Web. We are going to interface a current sensor to measure the energy consumption of the device that is connected to the project. The project will also be able to switch the device on and off remotely, and we are going to build an interface for you to control this switch from your computer and mobile device.

Chapter 3, Making Your Own Cloud-connected Camera, allows us to build our own DIY version of a wireless security camera by connecting a standard USB webcam to the Arduino Yún. We will perform two exciting applications with this project: first, we will automatically upload pictures from the camera when some motion is detected in front of it, and then we are going to make the camera stream video live from YouTube, so you can monitor what is going on in your home from anywhere.

Chapter 4, Wi-Fi-controlled Mobile Robot, focuses on robotics. We are going to build a Wi-Fi-controlled mobile robot with two wheels and an ultrasonic distance sensor in front of it. Additionally, we are going to use the powerful features of the Arduino Yún to easily control this robot via Wi-Fi. To do this, we are going to build a web interface that will be used to control the movement of the robot, and this will also display the distance measured by the front sensor.

What you need for this book

The main focus of this book is the Arduino Yún board; so, of course, you will need one of the Arduino Yún boards to make all four projects of the book. Depending on the chapter, you will also need several hardware components. The details of these components required are given at the beginning of each chapter.

You will also need to have some software installed on your computer to make the projects work. The first one is the Arduino IDE's latest beta version (the only version that can work with the Yún). For all these projects, I used the Arduino IDE Version 1.5.6-r2, but all the newer versions should work as well. You can download the Arduino IDE at http://arduino.cc/en/main/software#toc3.

You will also need a web server running on your computer for some of the projects. I recommend that you use a software that integrates a web server and a database, and that handles all the details for you. If you are using Windows, I recommend using EasyPHP, which is available at http://www.easyphp.org/.

Under OS X, I recommend using MAMP, which is available at http://www.mamp.info/.

With Linux, you can follow the instructions to install a web server provided at http://doc.ubuntu-fr.org/lamp.

Make sure the server is running at this point; we are going to use it in several of the projects in this book.

All the projects assume that your Arduino Yún board is already configured and connected to your Wi-Fi network. To configure and connect the Yún to your Wi-Fi network, there are only a few steps to follow. The first one is to plug the Arduino board in to the wall and wait for a moment.

After a while, you should see that a new Wi-Fi network has appeared in the list of Wi-Fi networks on your computer, created by the Yún. Connect to it, open a browser, and type the following command:

```
arduino.local
```

This should open a page served by the Arduino Yún board. You will be prompted to enter a password for your Yún board; please enter one that you can remember easily, as you will need it many times while attempting the projects in this book.

Then, you will be taken to a new page that contains some information about your Yún board. You can change the name of the board (which we will use later in all the projects), and also set your Wi-Fi parameters. You have to set these parameters so that the board can connect to your home Wi-Fi network. Choose the correct network from the list, enter your password, and click on **Configure & Restart**.

The Yún will then restart and connect to your network. At this point, you can also reconnect your computer to the local Wi-Fi network. After a while, you can type the following command in your browser along with the name you gave your Arduino board:

`myarduinoyun.local`

You should be taken to the same page again, but this time, with the Yún board connected to your local Wi-Fi network. If this is working, it means the Yún board is ready to be used for all the projects in the book.

You will also need to open the REST API of the Yún. This setting is configured on the configuration page of the Yún, where you have to select **OPEN**, which is close to **REST API ACCESS**. Reboot the Yún board again when the option has been changed.

Note that you have two ways to program your Yún board: you can either plug it directly into your computer via micro USB, or plug it into the wall via a USB adapter and upload the sketches via Wi-Fi.

Who this book is for

If you want to build exciting applications for the Internet of Things using the Arduino platform, this is the book for you. If you are planning to build some cool projects to automate your home and monitor it remotely, you will love this book. You will learn how to measure data, control devices, monitor your home remotely using a USB camera, and build a Wi-Fi-controlled mobile robot.

As far as skills are concerned, this book assumes that you already have some knowledge of the Arduino platform (for example, with the Arduino Uno) and some basic knowledge of electronics and programming. Note that the book can also be used without any previous experience with the Arduino Yún and the onboard Linux machine.

Conventions

In this book, you will find a number of styles of text that distinguish between different kinds of information. Here are some examples of these styles, and an explanation of their meaning.

Code words in text, database table names, folder names, filenames, file extensions, pathnames, dummy URLs, user input, and Twitter handles are shown as follows: "The alert mechanism occurs in the new function called `sendTempAlert` that is called if the temperature is below the limit."

A block of code is set as follows:

```
[default]
String data = "";
data = data + timeString + "," + String(temperature) + "," +
  String(humidity) + "," + String(lightLevel);
```

When we wish to draw your attention to a particular part of a code block, the relevant lines or items are set in bold:

```
[default]
if (client) {
  // Process request
  process(client);

  // Close connection and free resources.
  client.stop();
}
```

Any command-line input or output is written as follows:

```
# http://myarduinoyun.local/arduino/digital/8/1
```

New terms and **important words** are shown in bold. Words that you see on the screen, in menus or dialog boxes, for example, appear in the text like this: "Just click on **interface.html**, and the interface should open and be scaled to your phone screen size."

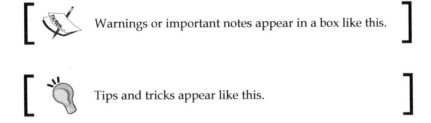

Warnings or important notes appear in a box like this.

Tips and tricks appear like this.

Reader feedback

Feedback from our readers is always welcome. Let us know what you think about this book—what you liked or may have disliked. Reader feedback is important for us to develop titles that you really get the most out of.

To send us general feedback, simply send an e-mail to feedback@packtpub.com, and mention the book title through the subject of your message.

If there is a topic that you have expertise in and you are interested in either writing or contributing to a book, see our author guide on www.packtpub.com/authors.

Customer support

Now that you are the proud owner of a Packt book, we have a number of things to help you to get the most from your purchase.

Downloading the example code

You can download the example code files for all Packt books you have purchased from your account at http://www.packtpub.com. If you purchased this book elsewhere, you can visit http://www.packtpub.com/support and register to have the files e-mailed directly to you.

All the up-to-date code for the four projects of this book can also be found at https://github.com/openhomeautomation/geeky-projects-yun.

Downloading the color images of the book

We also provide you a PDF file that has color images of the screenshots/diagrams used in this book. The color images will help you better understand the changes in the output. You can download this file from: https://www.packtpub.com/sites/default/files/downloads/8007OS_ColoredImages.pdf.

Errata

Although we have taken every care to ensure the accuracy of our content, mistakes do happen. If you find a mistake in one of our books—maybe a mistake in the text or the code—we would be grateful if you would report this to us. By doing so, you can save other readers from frustration and help us improve subsequent versions of this book. If you find any errata, please report them by visiting http://www.packtpub.com/support, selecting your book, clicking on the **errata submission form** link, and entering the details of your errata. Once your errata are verified, your submission will be accepted and the errata will be uploaded to our website, or added to any list of existing errata, under the Errata section of that title.

Piracy

Piracy of copyright material on the Internet is an ongoing problem across all media. At Packt, we take the protection of our copyright and licenses very seriously. If you come across any illegal copies of our works, in any form, on the Internet, please provide us with the location address or website name immediately so that we can pursue a remedy.

Please contact us at copyright@packtpub.com with a link to the suspected pirated material.

We appreciate your help in protecting our authors, and our ability to bring you valuable content.

Questions

You can contact us at questions@packtpub.com if you are having a problem with any aspect of the book, and we will do our best to address it.

1

Building a Weather Station Connected to the Cloud

This chapter will introduce you to the powerful features of the Arduino Yún microcontroller board. In this chapter, you will learn how to create a simple weather station that will send data to the cloud using the features of the web-based service **Temboo**. Temboo is not 100 percent free, but you will be able to make 1000 calls to Temboo per month using their free plan. You will learn how to connect sensors that measure temperature, humidity, and light level to your Arduino Yún. These sensors will first be separately tested to make sure that the hardware connections you made are correct.

Then, we are going to use the Temboo Arduino libraries to send these measurements to the cloud and to different web services so that they can be accessed remotely regardless of where you are in the world. Temboo is a web-based service that allows you to connect different web services together and proposes ready-to-use libraries for the Arduino Yún.

For example, the first thing we are going to do with Temboo is to send the data from your measurements to a Google Docs spreadsheet, where they will be logged along with the measurement data. Within this spreadsheet, you will be able to plot this data right in your web browser and see the data that arrives getting stored in your Google Docs account.

Then, we will use Temboo again to send an automated e-mail based on the recorded data. For example, you would like to send an alert when the temperature drops below a certain level in your home, indicating that a heater has to be turned on.

Finally, we will finish the chapter by using Temboo to post the data at regular intervals on a Twitter account, for example, every minute. By doing this, we can have a dedicated Twitter account for your home that different members of your family can follow to have live information about your home.

After completing this chapter, you'll be able to apply what you learned to other projects than just weather-related measurements. You can apply what you see in this chapter to any project that can measure data, in order to log this data on the Web and publish it on Twitter.

The Arduino Yún board is shown in the following image:

The required hardware and software components

Of course, you need to have your Arduino Yún board ready on your desk along with a micro USB cable to do the initial programming and testing. Also, we recommend that you have a power socket to the micro USB adapter so that you can power on your Arduino Yún directly from the wall without having your computer lying around. This will be useful at the end of the project, as you will want your Arduino Yún board to perform measurements autonomously.

You will then need the different sensors which will be used to sense data about the environment. For this project, we are going to use a DHT11 sensor to measure temperature and humidity and a simple photocell to measure light levels. DHT11 is a very cheap digital temperature and humidity sensor that is widely used with the Arduino platform. You can also use a DHT22 sensor, which is more precise, as the Arduino library is the same for both sensors. There are several manufacturers for these sensors, but you can find them easily, for example, on SparkFun or Adafruit. For the photocell, you can use any brand that you wish; it just needs to be a component that changes its resistance according to the intensity of the ambient light.

To make the DHT11 sensor and photocell work, we will need a 4.7k Ohm resistor and a 10k Ohm resistor as well. You will also need a small breadboard with at least two power rails on the side and some male-male jumper wires to make the electrical connections between the different components.

On the software side, you will need the latest beta version of the Arduino IDE, which is the only IDE that supports the Arduino Yún board (we used Version 1.5.5 when doing this project). You will also need the DHT library for the DHT11 sensor, which can be downloaded from `https://github.com/adafruit/DHT-sensor-library`.

To install the library, simply unzip the files and extract the `DHT` folder to your `libraries` folder in your main Arduino folder.

Connecting the sensors to the Arduino Yún board

Before doing anything related to the Web, we will first make sure that our hardware is working correctly. We are going to make the correct hardware connections between the different components and write a simple Arduino sketch to test all these sensors individually. By doing this, we will ensure that you make all the hardware connections correctly, and this will help a lot if you encounter problems in the next sections of this chapter that use more complex Arduino sketches.

The hardware connections required for our project are actually quite simple. We have to connect the DHT11 sensor and then the part responsible for the light level measurement with the photocell by performing the following steps:

1. First, we connect the Arduino Yún board's +5V pin to the red rail on the breadboard and the ground pin to the blue rail.

2. Then, we connect pin number 1 of the DHT11 sensor to the red rail on the breadboard and pin number 4 to the blue rail. Also, connect pin number 2 of the sensor to pin number 8 of the Arduino Yún board.

3. To complete the DHT11 sensor connections, clamp the 4.7k Ohm resistor between pin numbers 1 and 2 of the sensor.

For the photocell, first place the cell in series with the 10k Ohm resistor on the breadboard. This pull-down resistor will ensure that during the operation, if there is no light at all, the voltage seen by the Arduino board will be 0V. Then, connect the other end of the photocell to the red rail on the breadboard and the end of the resistor to the ground. Finally, connect the common pin to the Arduino Yún board analog pin A0.

The following image made using the Fritzing software summarizes the hardware connections:

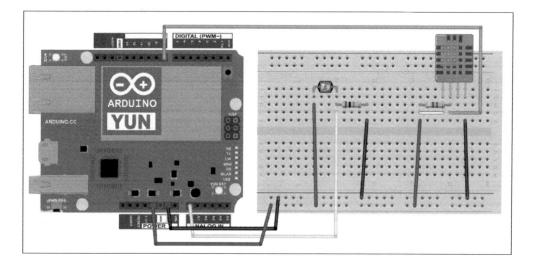

Now that the hardware connections are done, we will work on testing the sensors without uploading anything to the Web. Let's go through the important parts of the code.

First, we have to import the library for the DHT11 sensor, as follows:

```
#include "DHT.h"
```

Then, we need to declare a couple of variables that will store the measurements, as shown in the following code. These variables are declared as floats because the DHT sensor library returns float numbers.

```
int lightLevel;
float humidity;
float temperature;
```

Also, we can define the sensor pin and sensor type as follows:

```
#define DHTPIN 8
#define DHTTYPE DHT11
```

Create the DHT instance as follows:

```
DHT dht(DHTPIN, DHTTYPE);
```

Now, in the setup() part of the sketch, we need to start the serial connection, as follows:

```
Serial.begin(115200);
```

Next, in order to initialize the DHT sensor, we have the following:

```
dht.begin();
```

In the loop() part, we are going to perform the different measurements. First, we will calculate the temperature and humidity, as follows:

```
float humidity = dht.readHumidity();
float temperature = dht.readTemperature();
```

Then, measure the light level, as follows:

```
int lightLevel = analogRead(A0);
```

Finally, we print all the data on the serial monitor, as shown in the following code:

```
Serial.print("Temperature: ");
Serial.println(temperature);
Serial.print("Humidity: ");
Serial.println(humidity);
Serial.print("Light level: ");
Serial.println(lightLevel);
Serial.println("");
```

Repeat this every 2 seconds, as shown:

```
delay(2000);
```

The complete sketch for this part can be found at https://github.com/openhomeautomation/geeky-projects-yun/tree/master/chapter1/sensors_test.

Now it's time to test the sketch and upload it to the Arduino board. Then, open the serial monitor and you should have the data that comes from the sensors being displayed, as shown in the following screenshot:

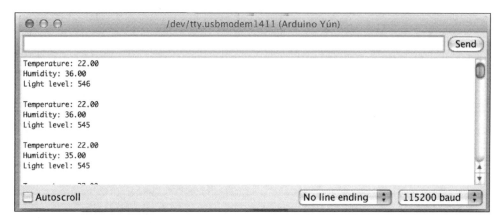

If you can see the different measurements being displayed as in the previous screenshot, it means that you have made the correct hardware connections on your breadboard and that you can proceed to the next sections of this chapter.

If it is not the case, please check all the connections again individually by following the instructions in this section. Please make sure that you haven't forgotten the 4.7k Ohm resistor with the DHT sensor, as the measurements from this sensor won't work without it.

Downloading the example code

You can download the example code files for all Packt books you have purchased from your account at http://www.packtpub.com. If you purchased this book elsewhere, you can visit http://www.packtpub.com/support and register to have the files e-mailed directly to you.

All the up-to-date code for the four projects of this book can also be found at https://github.com/openhomeautomation/geeky-projects-yun.

Creating a Temboo account

The next step in this project is to create and set up an account on the web service Temboo, so you can use the wide range of services provided by Temboo to upload data to Google Docs and to use their Gmail and Twitter libraries. This account will actually be used in the whole book for the other projects as well.

To do so, the first step is to simply go to the Temboo website at `http://temboo.com/`.

On the main page, simply enter your e-mail address to register and click on **Sign up**, as shown in the following screenshot:

You will then be asked to enter some basic information about your account, such as your account name, as shown in the following screenshot:

Then, you will be prompted to create your first app. Ensure that you save the details of your account, such as the name of your first app and the key that will be given to you; we are going to need it for the rest of this book.

In case you need to get these values again or create a new application, you can always access this data in the **My Account** section of the Temboo website by clicking on the **MANAGE** button below **APPLICATIONS**, just as it is displayed in the following screenshot:

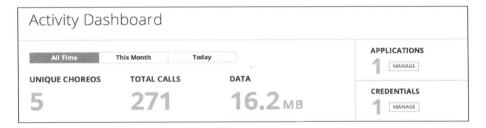

We are now all set to start using the Temboo libraries that are made specifically for the Arduino Yún board and to upload some data to the cloud.

Sending data to Google Docs and displaying it

In this section, we are going to use our first Temboo library (called a **Choreo**) to upload the measurements of the Arduino Yún to the Web and log the data into a Google Docs spreadsheet.

First, let's have a look at what a Choreo is and how you can generate the code for your Arduino Yún board. If you go to the main Temboo page, you will see that you can choose different platforms and languages, such as Arduino, JavaScript, or Python. Each of these links will allow you to select a Choreo, which is a dedicated library written for the platform you chose and can interface with a given web service such as Google Docs.

For the Arduino platform, Temboo even offers to generate the entire code for you. You can click on the Arduino icon on the Temboo website and then click on Arduino Yún; you will get access to a step-by-step interface to generate the code. However, as we want to get complete control of our device and write our own code, we won't use this feature for this project.

Google Docs is really convenient as it's an online (and free) version of the popular Office software from Microsoft. The main difference is that because it's all in the cloud, you don't have to store files locally or save them—it's all done online. For our project, the advantage is that you can access these files remotely from any web browser, even if you are not on your usual computer. You just need your Google account name and password and can access all your files.

If you don't have a Google account yet, you can create one in less than five minutes at `https://drive.google.com/`.

This will also create an account for the Gmail service, which we will also use later. Please make sure that you have your Google Docs username and password as you are going to need them soon.

Before we start writing any Arduino code, we need to prepare a Google Docs spreadsheet that will host the data. Simply create a new one at the root of your Google Docs account; you can name it whatever you wish (for example, `Yun`). This is done from the main page of Google Docs just by clicking on **Create**.

In the spreadsheet, you need to set the name of the columns for the data that will be logged; that is, Time, Temperature, Humidity, and Light level. This is shown in the following screenshot:

	A	B	C	D
1	**Time**	**Temperature**	**Humidity**	**Light level**
2				

Now, let's start building the Arduino sketch inside the Arduino IDE. We first need to import all the necessary libraries, as follows:

```
#include <Bridge.h>
#include <Temboo.h>
#include <Process.h>
```

The Bridge library is something that was introduced for the Arduino Yún board and is responsible for making the interface between the Linux machine of the Yún and the Atmel processor, where our Arduino sketch will run. With this library, it's possible to use the power of the Linux machine right inside the Arduino sketch.

The Process library will be used to run some programs on the Linux side, and the Temboo file will contain all the information that concerns your Temboo account. Please go inside this file to enter the information corresponding to your own account. This is as shown in the following code:

```
#define TEMBOO_ACCOUNT "temboo_account_name"  // Your Temboo
account name
#define TEMBOO_APP_KEY_NAME " temboo_app_name "  // Your Temboo
app key name
#define TEMBOO_APP_KEY " temboo_api_key "  // Your Temboo app key
```

> Note that we also included a debug mode in the sketch that you can set to true if you want some debug output to be printed on the serial monitor. However, for an autonomous operation of the board, we suggest that you disable this debugging mode to save some memory inside Yún.

In the sketch, we then have to enter the Google Docs information. You need to put your Google username and password here along with the name of the spreadsheet where you want the data to be logged, as shown in the following code:

```
const String GOOGLE_USERNAME = "yourUserName";
const String GOOGLE_PASSWORD = "yourPassword";
const String SPREADSHEET_TITLE = "Yun";
```

In the `setup()` part of the sketch, we are now starting the bridge between the Linux machine and the Atmel microcontroller by executing the following line of code:

```
Bridge.begin();
```

We are also starting a date process so that we can also log the data of when each measurement was recorded, as shown in the following code:

```
time = millis();
if (!date.running())  {
  date.begin("date");
  date.addParameter("+%D-%T");
  date.run();
}
```

The date will be in the format: date of the day followed by the time. The date process we are using here is actually a very common utility for Linux, and you can look for the documentation of this function on the Web to learn more about the different date and time formats that you can use.

Now, in the `loop()` part of the sketch, we send the measurements continuously using the following function:

```
runAppendRow(lightLevel, temperature, humidity);
```

Let's get into the details of this function. It starts by declaring the Choreo (the Temboo service) that we are going to use:

```
TembooChoreo AppendRowChoreo;
```

The preceding function is specific to Google Docs spreadsheets and works by sending a set of data separated by commas on a given row. There are Choreos for every service that Temboo connects to, such as Dropbox and Twitter. Please refer to the Temboo documentation pages to get the details about this specific Choreo. After declaring the Choreo, we have to add the different parameters of the Choreo as inputs. For example, the Google username, as shown in the following line of code:

```
AppendRowChoreo.addInput("Username", GOOGLE_USERNAME);
```

The same is done with the other required parameters, as shown in the following code:

```
AppendRowChoreo.addInput("Password", GOOGLE_PASSWORD);
AppendRowChoreo.addInput("SpreadsheetTitle", SPREADSHEET_TITLE);
```

The important part of the function is when we actually format the data so that it can be appended to the spreadsheet. Remember, the data needs to be delimited using commas so that it is appended to the correct columns in the spreadsheet, as shown in the following code:

```
String data = "";
data = data + timeString + "," + String(temperature) + "," +
String(humidity) + "," + String(lightLevel);
```

The Choreo is then executed with the following line of code:

```
unsigned int returnCode = AppendRowChoreo.run();
```

The function is then repeated every 10 minutes. Indeed, these values usually change slowly over the course of a day, so this is useless to the data that is logging continuously. Also, remember that the number of calls to Temboo is limited depending on the plan you chose (1000 calls per month on a free plan, which is approximately 1 call per hour). This is done using the delay function, as follows:

```
delay(600000);
```

For demonstration purposes, the data is logged every 10 minutes. However, you can change this just by changing the argument of the delay() function. The complete code for this part can be found at https://github.com/openhomeautomation/geeky-projects-yun/tree/master/chapter1/temboo_log.

You can now upload the sketch to the Arduino Yún board and open the Google Docs spreadsheet to see what's happening. It's all synchronized live with the Google Docs servers, so you do not need to refresh anything. After a while, you should see the first set of measurements being logged, as shown in the following screenshot:

	A	B	C	D
1	Time	Temperature	Humidity	Light level
2	02/28/14-11:57:23	28	33	638

In order to show you what can be done using this project, we used the integrated chart capabilities of Google Docs to plot this data using the measurements that we obtained for over 24 hours. The following screenshot is an extract from the raw data:

	A	B	C	D
1	**Time**	**Temperature**	**Humidity**	**Light level**
2	02/28/14-11:57:23	28	33	638
3	02/28/14-12:07:32	28	32	621
4	02/28/14-12:17:45	28	32	651
5	02/28/14-12:27:56	28	32	630
6	02/28/14-13:02:21	27	32	531
7	02/28/14-13:15:58	27	33	506
8	02/28/14-13:26:09	26	34	525
9	02/28/14-13:36:18	26	34	496
10	02/28/14-13:46:27	25	34	437
11	02/28/14-13:56:37	25	34	440
12	02/28/14-14:06:47	25	34	452
13	02/28/14-14:16:57	25	34	514

Now, to actually plot some data, you can simply use the **Insert charts** function of Google Docs. We chose the simple **Line** graph for our data. The following screenshot shows the results for temperature and humidity:

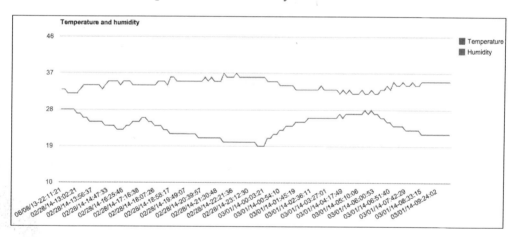

We did the same for light level measurements, as shown in the following screenshot:

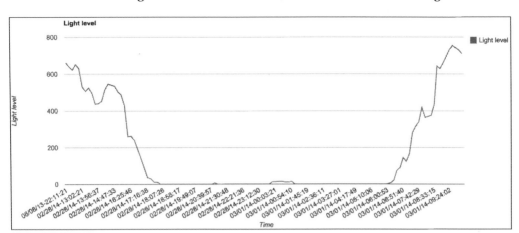

These charts can be placed automatically in their respective sheets inside your spreadsheet and will, of course, be updated automatically as new data comes in. You can also use the sharing capabilities of Google Docs to share these sheets with anyone, so they can also follow the measurements of your home.

Creating automated e-mail alerts

In this part, we are not only going to build on what we did in the previous section with Google Docs but also create some automated e-mail alerts on top with a Google account. This time, we will use the Temboo library that interfaces directly with Gmail, in this case, to automatically send an e-mail using your account.

What we will do is program the Arduino Yún board to send an e-mail to the chosen address if the temperature goes below a given level, for example, indicating that you should turn on the heating in your home.

Compared to the previous Arduino sketch, we need to add the destination e-mail address. I used my own address for testing purposes, but of course, this destination address can be completely different from the one of your Gmail account. For example, if you want to automatically e-mail somebody who is responsible for your home if something happens, execute the following line of code:

```
const String TO_EMAIL_ADDRESS = "your_email_address";
```

Please note that sending an e-mail to yourself might be seen as spam by your Gmail account. So, it's advisable to send these alerts to another e-mail of your choice, for example, on a dedicated account for these alerts. We also need to set a temperature limit in the sketch. In my version of the project, it is the temperature under which the Arduino Yún will send an e-mail alert, but you can of course modify the meaning of this temperature limit, as shown in the following line of code:

```
int temperature_limit = 25.0;
```

In the `loop()` part of the sketch, what changes compared to the sketch of the previous section is that we can compare the recorded temperature to the limit. This is done with a simple `if` statement:

```
if (temperature < temperature_limit) {
  if (debug_mode == true){Serial.println("Sending alert");}
    sendTempAlert("Temperature is too low!");
  }
```

Then, the alert mechanism occurs in the new function called `sendTempAlert` that is called if the temperature is below the limit. The function also takes a string as an argument, which is the content of the message that will be sent when the alert is triggered. Inside the function, we start again by declaring the type of Choreo that we will use. This time, the Choreo that we will use is specific to Gmail and is used to send an e-mail with the subject and body of the message, as shown in the following line of code:

```
TembooChoreo SendEmailChoreo;
```

Just as the Choreo we used to log data into Google Docs, this new Choreo requires a given set of parameters that are defined in the official Temboo documentation. We need to specify all the required inputs for the Choreo, for example, the e-mail's subject line that you can personalize as well, as shown in the following line of code:

```
SendEmailChoreo.addInput("Subject", "ALERT: Home Temperature");
```

The body of the message is defined in the following line of code:

```
SendEmailChoreo.addInput("MessageBody", message);
```

Note that the `message` variable is the one passed in the `loop()` part of the sketch and can be personalized as well, for example, by adding the value of the measured temperature. Finally, the Choreo is executed with the following line of code:

```
SendEmailChoreo.run();
```

The complete code for this part can be found at `https://github.com/ openhomeautomation/geeky-projects-yun/tree/master/chapter1/ temboo_alerts`.

Now, you can compile and update the sketch to your Yún. You can also go to the Gmail interface to check for new e-mails. If the temperature indeed drops below the value that you set as a limit, the following is what you should receive in your inbox:

Again, you can play with this sketch and create more complex alerts based on the data you measured. For example, you can add the humidity and light level in the mix and create dedicated limits and alerts for these values. You can also program Arduino Yún so that it e-mails you the data itself at regular intervals, even if no temperature limit is reached.

Making your Arduino Yún board tweet sensor data

Finally, in the last part of this project, we will make your Arduino Yún board send its own messages on Twitter. You can even create a new Twitter account just for your Yún board, and you can tell people you know to follow it on Twitter so that they can be informed at all times about what's going on in your home!

The project starts on the Twitter website because you have to declare a new app on Twitter. Log in using your Twitter credentials and then go to `https://apps.twitter.com/`.

Now, click on **Create New App** to start the process, as shown in the following screenshot:

You will need to give some name to your app. For example, we named ours MyYunTemboo. You will need to get a lot of information from the Twitter website. The first things you need to get are the API key and the API secret. These are available in the **API Keys** tab, as shown in the following screenshot:

Make sure that the **Access level** of your app is set to **Read**, **Write**, and **Direct** messages. This might not be active by default, and in the first tests, my Arduino board did not respond anymore because I didn't set these parameters correctly. So, make sure that your app has the correct access level.

Then, you are also going to need a token for your app. You can do this by going to the **Your access token** section. From this section, you need to get the **Access token** and the **Access token secret**. Again, make sure that the access level of your token is correctly set.

We can now proceed to write the Arduino sketch, so the Arduino Yún board can automatically send tweets. The Twitter Choreo is well known for using a lot of memory on the Yún, so this sketch will only tweet data without logging data into your Google Docs account. I also recommend that you disable any debugging messages on the serial port to preserve the memory of your Yún. In the sketch, you first need to define your Twitter app information, as shown in the following code:

```
const String TWITTER_ACCESS_TOKEN = "yourAccessToken";
const String TWITTER_ACCESS_TOKEN_SECRET = "
yourAccessTokenSecret";
const String TWITTER_API_KEY = " yourApiKey";
const String TWITTER_API_SECRET = " yourApiKeySecret";
```

Then, the sketch will regularly tweet the data about your home with the following function:

```
tweetAlert(lightLevel, temperature, humidity);
```

This function is repeated every minute using a delay() function, as follows:

```
delay(60000);
```

Of course, this delay can be changed according to your needs. Let's see the details of this function. It starts by declaring the correct Choreo to send updates on Twitter:

```
TembooChoreo StatusesUpdateChoreo;
```

Then, we build the text that we want to tweet as a string. In this case, we just formatted the sensor data in one string, as shown in the following code:

```
String tweetText = "Temperature: " + String(temperature) + ",
Humidity: " + String(humidity) + ", Light level: " +
String(light);
```

The access token and API key that we defined earlier are declared as inputs:

```
StatusesUpdateChoreo.addInput("AccessToken",
TWITTER_ACCESS_TOKEN);
StatusesUpdateChoreo.addInput("AccessTokenSecret",
TWITTER_ACCESS_TOKEN_SECRET);
StatusesUpdateChoreo.addInput("ConsumerKey", TWITTER_API_KEY);
StatusesUpdateChoreo.addInput("ConsumerSecret",
TWITTER_API_SECRET);
```

The text that we want to tweet is also simply declared as an input of the Twitter Choreo with the string variable we built earlier:

```
StatusesUpdateChoreo.addInput("StatusUpdate", tweetText);
```

The complete code for this part can be found at https://github.com/openhomeautomation/geeky-projects-yun/tree/master/chapter1/temboo_twitter.

Now that the Arduino sketch is ready, we can test it. You can simply upload the code to your Arduino Yún, and wait for a moment. Your board should automatically connect to the Twitter feed that you chose and print the data as a new message, as shown in the following screenshot:

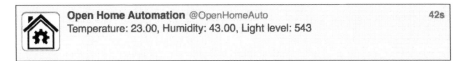

If nothing shows up on your Twitter account, there are several things that you can check. I already mentioned memory usage; try to disable the debug output on the serial port to free some memory. Also, make sure that you have entered the correct information about your Twitter app; it is quite easy to make a mistake between different API keys and access tokens.

For this project, I used the Twitter account of my website dedicated to home automation, but of course, you can create a dedicated Twitter account for the project so that many people can follow the latest updates about your home!

You can also combine the code from this part with the idea of the previous section, for example, to create automated alerts based on the measured data and post messages on Twitter accordingly.

Summary

Let's summarize what we learned in this chapter. We built a simple weather measurement station based on the Arduino Yún board that sends data automatically into the cloud.

First, you learned how to connect simple sensors to your Arduino Yún board and to write a test sketch for the Yún board in order to make sure that all these sensors are working correctly.

Then, we interfaced the Arduino Yún board to the Temboo services by using the dedicated Temboo libraries for the Yún. Using these libraries, we logged data in a Google Docs spreadsheet, created automated e-mail alerts based on our measurements, and published these measurements on Twitter.

To take it further, you can combine the different parts of this project together, and also add many Arduino Yún boards to the project, for example, in two different areas of your home. In the next chapter, we are going to use the power of the Temboo libraries again to send power measurement data to the Web, so the energy consumption of your home can be monitored remotely.

2
Creating a Remote Energy Monitoring and Control Device

In the second project of the book, we will continue to use the features of the Arduino Yún to connect to the Web using the web service Temboo. One thing people usually want to do in home automation is follow the energy consumption of their electrical devices and turn them on or off remotely, for example, using their smartphones or tablets.

Of course, many devices that currently exist can measure energy consumption on a power socket as well as being able to switch the device that is connected to this socket on and off. These devices are now very compact and easy to connect to a local Wi-Fi network, and these can also communicate with mobile devices using Bluetooth. Many large electronics manufacturers have developed their own solutions, and everyone can now buy these products and install them in their homes.

In this project, we are going to build our own do-it-yourself version of such a device and build a power switch and energy meter in order to turn an electrical device on and off as well as to follow its energy consumption.

The following are the main highlights of this chapter:

- Connecting a relay to one of the Arduino Yún digital outputs and using the Yún REST API to command this relay from a web browser
- Using an analog current sensor to get a measurement of the instant current consumption from the device that is connected to the relay, and calculate the instant power consumption from this measurement

- Sending this data to a Google Docs spreadsheet so that it can be accessed remotely from any web browser or from the Google Docs mobile app, and calculating the energy consumption and some other useful data such as the total energy cost of the device connected to your project
- Creating a simple web interface to control the lamp using your computer or any smartphone or tablet

The required hardware and software components

The first part of this project is to get the required parts that we are going to use for our energy consumption meter and power switch project. Apart from the Arduino Yún board, which will be the "brain" of the project, you will need to have two main parts ready on your desk when building the project. These parts are the relay module, which we will use to switch the lamp on and off, and the analog current sensor, which is used to measure the power and later the energy consumption of the lamp.

A relay is basically an electromagnetic switch used in projects where we need to switch a really large voltage (110V or 230V) using a small voltage as the command signal (5V from the Arduino board). For the relay, we used a basic 5V relay module from Polulu, which can switch up to 10A and is more than enough for many home appliances such as lamps. (In Europe, with 230V, you can connect up to 2300W.) The module itself is simply a relay mounted on a printed circuit board along with the required components that are necessary to operate the relay and some large headers and traces to carry up to 10A if necessary. It uses an Omron G5LE-14-DC5 relay. The following image is the relay used:

Of course, you can use any equivalent relay module. Just make sure that it can be switched on/off using a digital 5V signal like we have on the Arduino Yún board and that it can switch at least 5A, just to be safe for this project. The lamp we are using in this project only uses around 130 mA, but you may want to connect larger devices to your project later. If you want to build your own module from a relay, you simply need to add a diode in series with the relay to protect your Arduino board when the relay is switching.

Do not attempt to use a relay alone on a breadboard along with the required components to operate it. The small tracks on the breadboard cannot support high currents and voltages and you will run into serious safety issues if you do so, such as the potential meltdown of these tracks, which can lead to fire. So, use a dedicated relay module for this project.

Then, you need a current sensor to get the instant current consumption of the lamp. We used a module from ITead Studio, which is basically a breakout board for the ACS712 sensor. A breakout board is simply a board that is composed of a printed circuit board, the chip itself, and all the components required to make the chip work, such as resistors and capacitors. This sensor delivers an analog signal as an output, which is proportional to the measured current. This signal can then easily be converted to the corresponding current on the Arduino Yún board. We will acquire this analog signal using one of the integrated analog-digital converters of the Yún board. Note that there are also noninvasive current sensors that you can simply clip around the cable you want to measure, but these are usually bigger and don't integrate well with Arduino projects. The following is an image of the sensor used for this project:

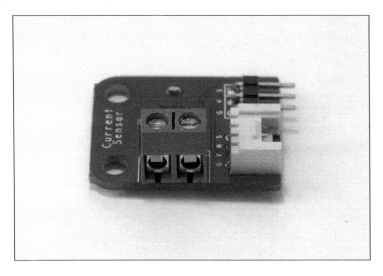

Just as for the relay module, you can use any equivalent current sensor for this project. The important parameters to be considered are the maximum current that can flow through the sensor (5A for the one we used) and the sensitivity of the sensor (185 mV/A for the one we used). If these two parameters are similar to the sensor I used in this project, or if they are better, you can use the sensor.

You also need to connect the lamp to the project in some way. Of course, one way would be to directly cut the power cable of the lamp and connect the lamp directly to our project, but I don't like this option because it's quite messy. As I mentioned in the introduction of this project, I don't want you to touch your lamp or other device in any way, and I want you to be able to connect your lamp again to the power socket in the wall if you want to.

I used two power cables so that I can connect my project to the wall socket on one side and connect the lamp to the project on the other side, just as I would do with a commercial device bought off the shelf.

The following is what I used for the power cable where I will plug the lamp:

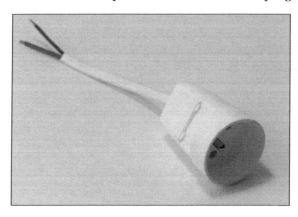

The following is the power cable I will use to connect the project to the wall plug:

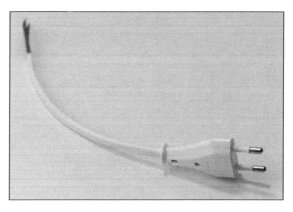

Note that some power plugs have three cables (the additional cable being for the ground connection), but as we will be working with small loads such as lamps, the third cable is not necessary for this project.

On the hardware side, I also used a small piece of a breadboard to make the power connections of the relay and the current sensor (because the Arduino Yún board only has one 5V pin). But of course, you can use a full-size breadboard to make these connections.

On the software side, you will need the latest beta version of the Arduino IDE, which is the only one that supports the Arduino Yún board (I used Version 1.5.5 while doing this project).

Connecting the components to the Yún board

We are now going to connect the relay module and the current sensor to the Arduino Yún board, connect the power cable that will power up the lamp, and finally connect everything to the power socket in the wall. This part is slightly more difficult than the hardware connections in *Chapter 1, Building a Weather Station Connected to the Cloud*, as it involves more steps and uses a higher voltage that requires you to take some precautions. So, please be careful and follow all the steps.

The first step is to put the Arduino Yún board, the relay module, and the current sensor board close to each other, as shown in the following image:

Then, we are going to connect the power supply of the relay module and the current sensor. As I said earlier in this chapter, the Arduino Yún board only has one 5V pin. This is why I connected the 5V pins of the two modules to a small piece of a breadboard first and then connected this breadboard to the Arduino 5V pin, as shown in the following image:

After this, we have to connect the ground pins of the two modules to the ground pin on the Arduino Yún board as shown in the following image. The Arduino Yún board has two ground pins on the board, so you don't have to use the breadboard for this.

To finish with the connection of the two modules, we need to connect their respective signal pins to the Arduino board. The relay will be controlled via pin number 8 of the Arduino Yún board, so connect the signal pin of the relay module to pin number 8 of the Yún board.

The current sensor has an analog output, so it has to be connected to one analog input on the Arduino board in order to acquire the signal using one of the Yún integrated analog-to-digital converters. This converter will acquire the analog signal that comes from the sensor and transform it into digital values that range from 0 to 1023 (which correspond to a 10-bit precision). Connect the output pin of the current sensor module to pin number A0 of the Arduino Yún board, as shown in the following image:

That's basically all for the low-power part. Now, we will focus on connecting the project to the two power supply cables so that we can plug the project into the wall plug and plug the lamp to the project. We will start by connecting the cable that will go to the wall, as shown in the following image:

Finally, connect the female power plug that we will connect the lamp to, as shown in the following image:

Finally, it's time to power up everything. You can plug your Arduino Yún board in to your computer via a USB cable (if you want to upload sketches directly and want space for your computer to be around the project) or via a wall power socket to a USB adapter (if you plan to upload the sketches via Wi-Fi and leave the project to work on its own).

Then, plug the lamp or the device that you want to control in to the female power plug of the project. To finish, connect the male power plug to the power socket in the wall. Be careful while performing this step: make sure that no electrical conductors are exposed, all screw terminals are correctly screwed and are holding the cables firmly, and no bare electrical conductors touch each other.

Testing your hardware connections

Now that the connections are done, we are going to test everything before we start sending energy consumption data to the cloud and building the interface to control the relay. We are going to test the different modules as if the project was already in operation. For the entire duration of the tests, we are going to connect the project to the power socket in the wall and to the lamp that we want to control. This way, we will ensure that all the hardware connections are correct before moving further.

The relay, for example, will be controlled via Wi-Fi using the Arduino Yún REST API, just as it will be in the final version of the project. Basically, we will just send a command from your web browser to directly set the value of the pin to which the relay is connected. Later in the project, we will make this call via a graphical user interface instead of actually typing the command in a browser.

For the current sensor, we are going to simply read the value measured on the analog pin A0 using the analog-to-digital converter of the Yún, convert it to a usable current, and then calculate the value of the effective current and the effective power as we already know the value of the effective voltage (110V or 230V depending on where you live).

Let's first have a look at the Arduino code. It starts by importing the right libraries, as shown in the following code. We need the `Bridge` library so that we can use the functions from the onboard Linux machine of the Yún, and the `YunServer` and `YunClient` libraries so that we can receive external commands using the REST API. REST APIs are usually only used by web developers, but Arduino actually proposes a sketch that implements such an API for the Arduino Yún. This sketch is directly accessible in the library that comes with the Arduino Yún, and in this project, I used a modified version of this reference sketch.

```
#include <Bridge.h>
#include <YunServer.h>
#include <YunClient.h>
```

To use the REST API of the Yún, we need to create a `YunServer` instance, as shown in the following line of code. This server will run continuously and wait for incoming commands.

```
YunServer server;
```

We also need to define the pins that our sensors are connected to, as shown in the following lines of code:

```
#define CURRENT_SENSOR A0
#define RELAY_PIN 8
```

One important part of the sketch is to declare the value of the effective voltage, which will be used later to calculate the effective power of the device, as shown in the following line of code. This value depends on where you are located (for example, 230 for Europe, and 110 for the USA):

```
float effective_voltage = 230;
```

In the `setup()` part, we need to start the bridge between the Arduino microcontroller and the Linux machine as follows:

```
Bridge.begin();
```

We also need to start the web server as follows:

```
server.listenOnLocalhost();
server.begin();
```

Then, the last and most important part of the `setup()` function is to calibrate the sensor in order to determine which value is returned when the current is null. This is done by the following line of code:

```
zero_sensor = getSensorValue();
```

Let's dive into this function. We could simply get one measurement from the current sensor but that would be a bad idea. Indeed, the value that you get from the sensor varies slightly over time, around an average that we actually want to measure. This is typical behavior when using analog sensors that have important sensitivities such as the one we are using here. This is why the function basically samples and averages the signal over several measurements with the following lines of code:

```
for (int i = 0; i < nb_measurements; i++) {
  sensorValue = analogRead(CURRENT_SENSOR);
  avgSensor = avgSensor + float(sensorValue);
}
avgSensor = avgSensor/float(nb_measurements);
```

After these measurements, we return the average as follows:

```
return avgSensor;
```

This way, we are sure to get a stable value of the sensor reading every time. This value is then used throughout the whole sketch as a reference value for the current sensor readings. For example, if the measured value is equal to this reference, we will know that the current in the lamp is null. The actual sensor reading during the operation of the project uses the same function, so we always get a stable measurement.

Now, comes the loop() part of the sketch. It actually consists of two parts: in the first part, we will receive incoming connections on the Yún web server that we started earlier, and in the second part, we will print out the measurements that come from the current sensor.

For the web server part, we can listen for connections as follows:

```
YunClient client = server.accept();
```

If a client is detected, we process the request with the following code:

```
if (client) {
  // Process request
  process(client);

  // Close connection and free resources.
  client.stop();
}
```

I won't detail the Process function as it is the same as in the Bridge example for the Arduino Yún that we used earlier (this is available as an example in the Yún Bridge library). To know more about the Yún REST API, you can visit the official Arduino documentation on the Arduino website at http://arduino.cc/en/Guide/ArduinoYun.

Now, we will write the part of the sketch that is responsible for the current measurements. It starts when you get a stable measurement, just as we did earlier for the null current as follows:

```
float sensor_value = getSensorValue();
```

I won't get into the details of this function as it is the same as which we used to get the value for the null current. We can now do some calculations on this measured value. First, we need to convert it to a usable current value as follows:

```
amplitude_current=(float)(sensor_value-
zero_sensor)/1024*5/185*1000000;
```

This is the amplitude of the current, which is a sinusoidal current. This formula can be found in the datasheet of the sensor as well as on the ITead Studio website. Because we know this information about the current, to get the effective current, we simply need to divide it by the square root of two as follows:

```
effective_value=amplitude_current/1.414;
```

To get the effective power, we then need to transform this current in amperes by dividing the value by 1000 and multiplying it with the effective voltage. I also added an absolute value operator so that the power is always positive, even when you connect the current sensor to measure negative currents, as follows:

```
abs(effective_value*effective_voltage/1000);
```

The sketch ends by printing all these values on the Serial monitor and repeats itself every 50 milliseconds. The complete sketch for this part is available on the GitHub repository of the book at `https://github.com/openhomeautomation/geeky-projects-yun/tree/master/chapter2/sensors_test`.

Now you can upload the sketch to the Arduino board. Remember that at this point, the Arduino Yún board should be powered by either your computer or a USB power adapter, the lamp should be plugged to the project in the female power cord, and the project itself should be plugged into the wall socket.

The relay is quite easy to test; you just need to go to your web browser and type in the right command. The REST API of the Yún works by typing the name of your Arduino Yún board followed by `.local` (in my case, I named it `myarduinoyun`). Then, if it is followed by `arduino/`, you can directly use commands to change the value of the Arduino pins. For example, to change the value of the relay pin to 1, you need to add `digital/8/1`, as shown in the following screenshot:

The preceding command means that you are calling the command `digitalWrite(8,HIGH)` using the REST API. You should instantly hear the relay switch and see the light turn on. Try again by adding a `0` after the command instead of a `1`; the relay should switch again and turn the light off. Don't worry, as later in the project, we'll build a nice graphical interface so that you don't have to write this command every time.

Now we are going to check the measurements coming from the current sensor. Make sure that the lamp is off, reset the Arduino microcontroller to be sure that the sketch starts from the beginning again, and then open the Serial monitor. To do this, the Arduino Yún board has to be connected to your computer via the USB cable. The first thing you should see is the measurement for a null current as follows:

```
Sensor value: 493.05
```

Then, the sketch continuously displays the value of the sensor reading, current amplitude, effective current, and effective power. Even if the current is null, remember that we average the sensor readings over several measurements, so there can be minor fluctuations in the value, as shown in the following code:

```
Sensor value: 492.87
Current amplitude (in mA):
4.5
Current effective value (in mA)
3.2
Effective power (in W):
0.7
```

If you then turn the lamp on using the REST call in your browser, you should instantly see a change in the current and power readings as follows:

```
Sensor value: 486.52
Current amplitude (in mA):
-163.1
Current effective value (in mA)
-115.4
Effective power (in W):
26.5
```

If you can see these values and your relay is responding to the REST calls in your browser, it means that your hardware is working correctly and you can proceed to the next step. If it doesn't work, the first step is to check the different connections of the current sensor and relay module. Also check that you have selected the correct Serial speed in the Serial monitor so that it matches the speed defined in the sketch.

Sending data to Google Docs

The first step is to set up a Google Docs spreadsheet for the project. Create a new sheet, give it a name (I named mine `Power` for this project, but you can name it as you wish), and set a title for the columns that we are going to use: **Time**, **Interval**, **Power**, and **Energy** (that will be calculated from the first two columns), as shown in the following screenshot:

We can also calculate the value of the energy using the other measurements. From theory, we know that over a given period of time, energy is power multiplied by time; that is, *Energy = Power * Time*.

However, in our case, power is calculated at regular intervals, and we want to estimate the energy consumption for each of these intervals. In mathematical terms, this means we need to calculate the integral of power as a function of time.

We don't have the exact function between time and power as we sample this function at regular time intervals, but we can estimate this integral using a method called the trapezoidal rule. It means that we basically estimate the integral of the function, which is the area below the power curve, by a trapeze. The energy in the c2 cell in the spreadsheet is then given by the formula:

Energy= (PowerMeasurement + NextPowerMeasurement)*TimeInverval/2.

Concretely, in Google Docs, you will need the formula, *D2 = (B2 + B3)*C2/2*.

The Arduino Yún board will give you the power measurement, and the time interval is given by the value we set in the sketch. However, the time between two measurements can vary from measurement to measurement. This is due to the delay introduced by the network. To solve this issue, we will transmit the exact value along with the power measurement to get a much better estimate of the energy consumption.

Then, it's time to build the sketch that we will use for the project. The goal of this sketch is basically to wait for commands that come from the network, to switch the relay on or off, and to send data to the Google Docs spreadsheet at regular intervals to keep track of the energy consumption.

We will build the sketch on top of the sketch we built earlier so I will explain which components need to be added. First, you need to include your Temboo credentials using the following line of code:

```
#include "TembooAccount.h"
```

Since we can't continuously measure the power consumption data (the data transmitted would be huge, and we will quickly exceed our monthly access limit for Temboo!), like in the test sketch, we need to measure it at given intervals only. However, at the same time, we need to continuously check whether a command is received from the outside to switch the state of the relay. This is done by setting the correct timings first, as shown in the following code:

```
int server_poll_time = 50;
int power_measurement_delay = 10000;
int power_measurement_cycles_max = power_measurement_delay/server_
poll_time;
```

The server poll time will be the interval at which we check the incoming connections. The power measurement delay, as you can guess, is the delay at which the power is measured.

However, we can't use a simple delay function for this as it will put the entire sketch on hold. What we are going to do instead is to count the number of cycles of the main loop and then trigger a measurement when the right amount of cycles have been reached using a simple `if` statement. The right amount of cycles is given by the power measurement `cycles_max` variable.

You also need to insert your Google Docs credentials using the following lines of code:

```
const String GOOGLE_USERNAME = "yourGoogleUsername";
const String GOOGLE_PASSWORD = "yourGooglePass";
const String SPREADSHEET_TITLE = "Power";
```

In the `setup()` function, you need to start a date process that will keep a track of the measurement date. We want to keep a track of the measurement over several days, so we will transmit the date of the day as well as the time, as shown in the following code:

```
time = millis();
if (!date.running())  {
  date.begin("date");
  date.addParameter("+%D-%T");
  date.run();
}
```

In the `loop()` function of the sketch, we check whether it's time to perform a measurement from the current sensor, as shown in the following line of code:

```
if (power_measurement_cycles > power_measurement_cycles_max)
```

If that's the case, we measure the sensor value, as follows:

```
float sensor_value = getSensorValue();
```

We also get the exact measurement interval that we will transmit along with the measured power to get a correct estimate of the energy consumption, as follows:

```
measurements_interval = millis() - last_measurement;
last_measurement = millis();
```

We then calculate the effective power from the data we already have. The amplitude of the current is obtained from the sensor measurements as shown earlier. Then we can get the effective value of the current by dividing this amplitude by the square root of 2. Finally, as we know the effective voltage and that power is current multiplied by voltage, we can calculate the effective power as well, as shown in the following code:

```
// Convert to current
amplitude_current=(float)(sensor_value-zero_
sensor)/1024*5/185*1000000;
effective_value=amplitude_current/1.414;

// Calculate power
    float effective_power = abs(effective_value *
effective_voltage/1000);
```

After this, we send the data with the time interval to Google Docs and reset the counter for power measurements, as follows:

```
runAppendRow(measurements_interval,effective_power);
power_measurement_cycles = 0;
```

The function to send data to Google Docs is nearly the same as the one we saw in *Chapter 1, Building a Weather Station Connected to the Cloud*. Let's quickly go into the details of this function. It starts by declaring the type of Temboo library we want to use, as follows:

```
TembooChoreo AppendRowChoreo;
```

Start with the following line of code:

```
AppendRowChoreo.begin();
```

We then need to set the data that concerns your Google account, for example, the username, as follows:

```
AppendRowChoreo.addInput("Username", GOOGLE_USERNAME);
```

The actual formatting of the data is done with the following line of code:

```
data = data + timeString + "," + String(interval) + "," +
String(effectiveValue);
```

Here, `interval` is the time interval between two measurements, and `effectiveValue` is the value of the measured power that we want to log on to Google Docs. The Choreo is then executed with the following line of code:

```
AppendRowChoreo.run();
```

Finally, we do this after every 50 milliseconds and get an increment to the power measurement counter each time, as follows:

```
delay(server_poll_time);
power_measurement_cycles++;
```

The complete code for this section is available at `https://github.com/ openhomeautomation/geeky-projects-yun/tree/master/chapter2/energy_log`.

The code for this part is complete. You can now upload the sketch and after that, open the Google Docs spreadsheet and then just wait until the first measurement arrives. The following screenshot shows the first measurement I got:

	A	B	C	D
1	Time	Interval	Power	Energy
2	03/12/14-11:12:01	19285	0.09	0.867825

After a few moments, I got several measurements logged on my Google Docs spreadsheet. I also played a bit with the lamp control by switching it on and off so that we can actually see changes in the measured data. The following screenshot shows the first few measurements:

	A	B	C	D
1	Time	Interval	Power	Energy
2	03/12/14-11:12:01	19285	0.09	2.12135
3	03/12/14-11:12:22	20732	0.13	6.2196
4	03/12/14-11:12:44	21473	0.47	12.45434
5	03/12/14-11:13:03	19153	0.69	13.98169
6	03/12/14-11:13:22	18858	0.77	11.69196
7	03/12/14-11:13:41	19205	0.47	8.25815
8	03/12/14-11:14:01	20262	0.39	6.0786

It's good to have some data logged in the spreadsheet, but it is even better to display this data in a graph. I used the built-in plotting capabilities of Google Docs to plot the power consumption over time on a graph, as shown in the following screenshot:

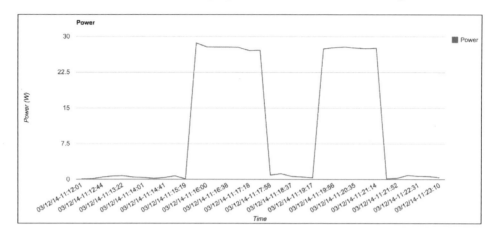

Using the same kind of graph, you can also plot the calculated energy consumption data over time, as shown in the following screenshot:

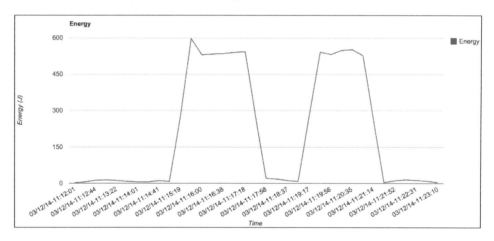

From the data you get in this Google Docs spreadsheet, it is also quite easy to get other interesting data. You can, for example, estimate the total energy consumption over time and the price that it will cost you. The first step is to calculate the sum of the energy consumption column using the integrated sum functionality of Google Docs.

Then, you have the energy consumption in Joules, but that's not what the electricity company usually charges you for. Instead, they use kWh, which is basically the Joule value divided by 3,600,000. The last thing we need is the price of a single kWh. Of course, this will depend on the country you're living in, but at the time of writing this book, the price in the USA was approximately $0.16 per kWh. To get the total price, you then just need to multiply the total energy consumption in kWh with the price per kWh.

This is the result with the data I recorded. Of course, as I only took a short sample of data, it cost me nearly nothing in the end, as shown in the following screenshot:

Energy consumption (J)	7,267.2931	
Energy consumption (kWh)	0.0020	
Price/kWh ($)	0.16	
Total price ($)	0.0003	

You can also estimate the on/off time of the device you are measuring. For this purpose, I simply added an additional column next to Energy named On/Off. I simply used the formula =IF(C2<2;0;1).

It means that if the power is less than 2W, we count it as an off state; otherwise, we count it as an on state. I didn't set the condition to 0W to count it as an off state because of the small fluctuations over time from the current sensor. Then, when you have this data about the different on/off states, it's quite simple to count the number of occurrences of each state, for example, on states, using =COUNTIF(E:E,"1").

I applied these formulas in my Google Docs spreadsheet, and the following screenshot is the result with the sample data I recorded:

On time	13	
Off time	22	

It is also very convenient to represent this data in a graph. For this, I used a pie chart, which I believe is the most adaptable graph for this kind of data. The following screenshot is what I got with my measurements:

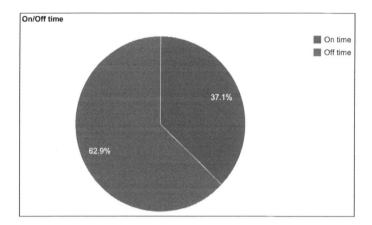

With the preceding kind of chart, you can compare the usage of a given lamp from day to day, for example, to know whether you have left the lights on when you are not there.

Building an interface to switch the lights on/off remotely

Now that our project automatically logs data concerning the energy consumption on Google Docs, it's time to go back to the relay control. For now, we tested the relay by going into a web browser and typing the correct REST function with the name of the pin we want to change.

However, that's not very convenient. You don't want to type something in your web browser every time you want to turn a light on in your home. What we would like to have instead is a nice graphical interface with buttons that can be pressed to turn a light on or off. It would be even better if this interface could be accessed not only from a web browser on your computer but also from any smartphone or tablet in your home. That's exactly what we are going to build now.

We need several components to do so, and we will mix several programming languages to build the best graphical interface possible. We will use HTML for the main page that will host the on/off button, JavaScript to handle the actions of this button, and PHP to transmit the correct command to the Arduino server. We are also going to use some CSS to make the interface look much better and automatically adapt itself to the device you are using, such as a smartphone.

First, let's deal with the HTML code. We need to import the jQuery library and the file that will contain all the JavaScript code, as follows:

```
<script src="jquery-2.0.3.min.js"></script>
<script src="script.js"></script>
```

Also, import the CSS style file, as follows:

```
<link rel="stylesheet" type="text/css" href="style.css">
```

The core of this HTML file is to create two buttons; one button to switch the relay on and the other to switch it off again. The following, for example, is the code for the **On** button:

```
<input type="button" id="on" class="commandButton" value="On"
onClick="relayOn()"/>
```

Now, if you were to actually take this file as it is, it would look really bad as some default styles would be applied to the buttons. That's why we attached a CSS file to make the interface look a bit better. For example, I decided to center align the main form that contains the two buttons, as follows:

```
#relay {
   text-align: center;
}
```

I also gave some style to the buttons themselves, such as an orange background; I made them bigger and also put a nice black border around them, as shown in the following code:

```
.commandButton {
   background-color: orange;
   border: 1px solid black;
   font-size: 40px;
   cursor: pointer;
   border-radius: 10px;
   width: 300px;
   height: 100px;
}
```

Now the interface looks much better on your computer. But what if somebody opens it from a smartphone? It would not be adapted at all to the tiny screen of a smartphone. To automatically make the interface adapt to the device you are using, we will use a property from CSS3 called media queries. This feature of CSS3 can, for example, detect whether a smaller screen size is used to access the page. Then, when you have this information, you can use it to modify the style of the different elements accordingly, for example, we may want to make our buttons appear differently on the page.

In our case, we want to make the buttons take all the space available on the smaller screen. We also want to double the height of each button as well as double the font size so that they can be really readable on a small screen like on a smartphone. All of this is done by the following piece of code:

```
@media screen and (max-device-width: 400px) {
  .commandButton {
    width: 100%;
    height: 200px;
    border: 2px solid black;
    font-size: 80px;
    margin-bottom: 50px;
      text-align: center;
      background-color: orange;
      }
}
```

The JavaScript file simply makes the interface between the GUI we just designed and the PHP file that will actually connect to the Arduino Yún board. The following, for example, is the code called by one button:

```
function relayOn(){
  $.get( "update_state.php", { command: "1"} );
}
```

The command variable simply contains the state of the relay that we want to send to the Arduino Yún board and will set the value of the pin that the relay is connected to.

Now let's see the PHP file. The first line of the code gets the command variable from the JavaScript and builds the command that will be sent to the Yún, as follows:

```
$service_url = 'http://myarduinoyun.local/arduino/digital/8/' . $_
GET["command"];
```

To actually send the command, we are going to use a PHP function named curl that we will use to call the REST API of the Yún. We first have to initialize this function with the URL we built earlier, as follows:

```
$curl = curl_init($service_url);
```

Finally, we actually execute this command with the following code:

```
curl_setopt($curl, CURLOPT_IPRESOLVE, CURL_IPRESOLVE_V4 );
$curl_response = curl_exec($curl);
curl_close($curl);
```

The option with `set` in the first line of code is used simply to speed up access to the Arduino board. Before testing the interface, make sure that the web server on your computer is running and that all the files of the project are located at the root of the web server folder. The complete code for this part of the project is available at `https://github.com/openhomeautomation/geeky-projects-yun/tree/master/chapter2/web_interface`.

You should see the two buttons of the interface show up in your browser, as shown in the following screenshot:

You can now test this simple interface. Just click on a button and the PHP code should give the correct command on your Arduino Yún board, making the switch go on or off instantly.

You can also test the interface on a smartphone or tablet. I used my phone to do so. Just open your favorite browser, go to your computer's IP address or network name, and you should see the different files of your project being displayed. Just click on **interface.html** and the interface should open and scale to your phone's screen size, as shown in the following screenshot:

Just as for the interface on your computer, you can simply press a button and the light will switch on or off instantly. Now, you are able to command this light from wherever you are in your home; you just have to be connected to your local Wi-Fi network.

Summary

Let's see what we learned in this project. At the beginning of the project, you saw how to interface the required components of this project to your Arduino Yún board: a relay module, a current sensor, and a lamp that will be controlled by the Yún board.

Then, we wrote a simple sketch to test the different components of the project and made sure that they all worked correctly.

Then, we built the energy consumption logging part of the project, and logged the power consumption inside a Google Docs spreadsheet. We also used the built-in capabilities of Google Docs to calculate the actual energy consumption, total energy cost, and on/off time of the device.

Finally, in the last part of the project, we built a graphical user interface to control the relay from a web browser, from your computer, or a smartphone/tablet.

Of course, there are many ways to take what you've learned in this project and extend it further. The first thing you can do is to add more devices to the project. For example, Arduino Yún has six analog inputs in total, so in theory you could plug the same number of current sensors into the Yún. Following the same principles, you could also add more Arduino Yún boards to the project.

You could also use the project with more features of Temboo, such as integrating the power measurements with social media, for example, by alerting the user with Twitter when the power consumption exceeds a given threshold. The user could then shut the lamp off by replying to this tweet.

In the next chapter, we will use other features of the Arduino Yún such as the USB port and the embedded Linux machine to create a wireless security camera. This camera will automatically upload pictures to a Dropbox folder and also stream the video live on YouTube so you can monitor your home remotely.

3

Making Your Own Cloud-connected Camera

In this project, we are going to build a security camera that automatically uploads pictures to the Web. We will connect a camera to the Arduino Yún board, and use its powerful features to control this camera easily and upload pictures to the Web. What we are going to build is a system that can detect motion, and if some motion is detected, can automatically take a picture and save it both on the local SD card attached to the Yún board and to a cloud storage; in our case, Dropbox. We are also going to make the camera stream a video live on a private YouTube feed.

Getting started

Let's see what we are going to do in this project in more detail:

- First, we are going to build the hardware part of the project with a typical USB camera, a PIR motion sensor, and one SD card.

- Then, we will write some code to test all the hardware connections of the project. We'll check whether the motion sensor is working correctly and try to take a picture with the camera while it is connected to the Arduino Yún board.

- After testing the hardware, we are going to build the first application, which captures pictures whenever some motion is detected and automatically stores these pictures on the SD card.

- Right after building this simple local application, we are going to connect the project to the cloud. The project will do the same as in the earlier case, take pictures when some motion is detected, but this time the pictures will also be uploaded to your Dropbox folder. This way, the pictures can be seen in real time from anywhere, as you can log in to Dropbox from any web browser.

- Finally, we are going to stream a video to the Web, so you can always check what's going on in your home from a mobile phone or tablet, wherever you are in the world. For this application, we are going to install a streaming library on the Yún board and make it continuously stream a video over Wi-Fi. This stream will be acquired by your computer and sent to YouTube via a dedicated software. On YouTube, we will then be able to access this live stream just as you would watch a typical YouTube video.

The required hardware and software components

First, let's see which components we need for the project. Apart from the Yún board, you will need three components: a USB camera, a PIR motion sensor, and an SD card. We will only make direct connections to Yún in this part, so you won't need a breadboard to make electrical connections.

The most important component of this project is the USB camera. We are using a standard USB webcam from Logitech, the C700 model, which can record pictures up to the HD resolution. Of course, you can use other cameras if you already have one on your desk. Make sure that the camera is compatible with **USB Video Class (UVC)**. Most of the recent webcams are compatible with this protocol. It might work with a camera that is not officially compatible with UVC, but there is no guarantee. You can find a list of all UVC compatible cameras at `http://en.wikipedia.org/wiki/List_of_USB_video_class_devices`.

Also, try to choose a camera with at least HD resolution, so you can get nice and clear pictures. It's not so important for the streaming part, but can be great if you want to use this project for other applications than security, for example, to create time-lapse videos. The following is an image of the USB camera we are using, the C700 USB webcam from Logitech:

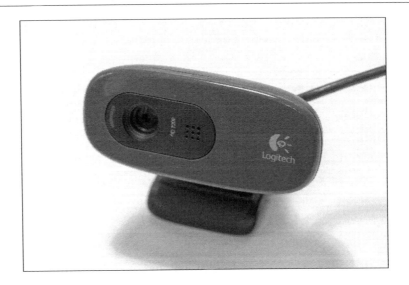

Then, there is the PIR motion sensor. This sensor is a really inexpensive sensor that uses infrared pictures to detect motion in a room from anything that emits heat, such as humans. We could have used the camera directly to detect motion, but that would have not been so efficient. The camera uses quite a lot of power when it is on, whereas a PIR motion sensor uses nearly no power. It would also have been more difficult to write the software required to detect motion efficiently from the camera recording. We used a PIR motion sensor from Parallax, which you can see in the following image:

Again, you can use other brands of PIR sensors. The main thing to consider is that it should work with 5V voltage levels because that is the voltage level used by the Yún. Most sensors work with both 3.3V and 5V voltage levels, so you shouldn't have many problems with this characteristic. When motion is detected, it should simply put a logical high level on its signal pin.

For the SD card, we used a standard micro SD card. Usually, you will have one already in your digital camera or smartphone. You will need to format it correctly so that the Yún can use it. We recommend that you use the official SD card formatter from the SD card association, see `https://www.sdcard.org/downloads/formatter_4/`.

Now, on the software side, you will need a bit more than just the Arduino IDE. We are going to install the required software for the camera directly on the Yún board when we connect to it via SSH, but you will need the Temboo Python SDK to upload pictures on to Dropbox. You can find the SDK at `https://www.temboo.com/download`.

Then, you also need to have a Dropbox account, so you can upload pictures on to it. You can simply create an account by going to `https://www.dropbox.com/home`.

Once your account is created, you need to create an app that will be used by your project. This basically means that you have to authorize the project you are going to build in this chapter to automatically send pictures to your Dropbox account without having to enter your login and password every time. You will also be given all the required information (such as an API key) that we will enter later in the Python script on Yún.

Perform the following steps to create an app:

1. To create an app, first go to `https://www.dropbox.com/developers/apps`.

2. Then, click on **Create app** in the top-right corner of the window. You can now choose the type of app you want to create. In our case, we want to use the **Dropbox API** directly, as shown in the following screenshot:

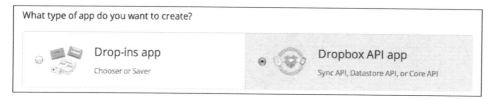

3. You will then be prompted to choose the kind of data your app needs to store. We want to upload pictures, so choose **Files and datastores**, as shown in the following screenshot:

What type of data does your app need to store on Dropbox?

◉ Files and datastores

○ Datastores only

4. You can then finish the process of creating your Dropbox app.

5. On the confirmation page that describes the app, you will need to write down the **App key** and **App secret**, which we will need for the rest of the project.

6. Also, make sure that the **Permission type** field is set to **App folder**. This will ensure that the pictures are uploaded to the folder dedicated to the app and that the Yún won't have access to the rest of your Dropbox folder.

7. What you need to get now is the Token key and Token secret relative to your Dropbox app, so you can enter them later in the software of our project.

   To get them, the first step is to go to the InitialiseOAuth Choreo on the Temboo website at `https://temboo.com/library/Library/Dropbox/OAuth/InitializeOAuth/`. Here, you will need to enter the App key and App secret. This will generate some additional information such as a callback ID and a temporary token secret. You'll also be asked to visit a link to Dropbox to confirm the authentication.

8. Finally, go to the FinalizeOAuth page to finish the process. You'll be asked to enter your App key, App secret, callback ID, and temporary token secret at `https://temboo.com/library/Library/Dropbox/OAuth/FinalizeOAuth/`.

   After this step, you'll be given your final Token key and Token secret. Write them down as you'll need them later.

Making hardware connections

It's now time to assemble our project. As we are going to use most of the Yún's connectivity, such as the USB port, it will be quite easy and quick to assemble the project. First, simply put the formatted micro SD card into the SD card reader of the Yún, which is located below the Yún board, as shown in the following image:

Then, plug the USB camera into the Yún USB port, as shown:

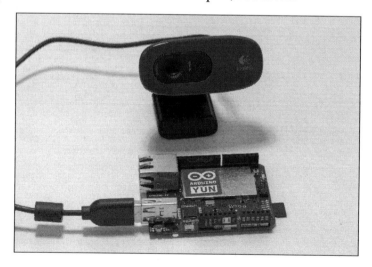

Finally, you need to connect the PIR motion sensor to the Yún board. It basically has three pins: VCC, GND, and SIG (signal pin). Connect VCC to the Yún's 5V pin, GND to the Yún ground, and SIG to pin number 8 of the Yún. You should end up with a setup similar to the following image:

Finally, you can connect the Yún to your computer via a micro USB cable or power it with a USB adapter if you want to use the project remotely and upload the Arduino sketches via Wi-Fi.

Testing your hardware connections

Now that all the connections are made, we can test the project. To get started, we will take care of the motion sensor. For this, we will write a very simple sketch that will only make use of the embedded Atmel microcontroller on the Yún board. We first need to declare the pin that the sensor is connected to, as follows:

```
const int sensor_pin = 8;
```

Then, in the `setup()` function, we will start the Serial connection, as follows:

```
Serial.begin(9600);
delay(1000);
```

We can also set some delay before data is read from the sensor, as it needs some time to initialize and work correctly. In the `loop()` function, we continuously read the value from pin number 8. Remember that the sensor will simply return a logical high state if some motion is detected and a low state otherwise. This means that we can store the sensor's reading into a Boolean variable, as shown in the following line of code:

```
boolean sensor_value = digitalRead(sensor_pin);
```

Every second, this value is then printed on the Serial monitor using the following lines of code:

```
Serial.println(sensor_value);
delay(100);
```

The complete code for this part can be found at `https://github.com/openhomeautomation/geeky-projects-yun/tree/master/chapter3/pir_test`.

You can now upload the preceding code on to your Yún board. Open the Serial monitor and try to pass your hand in front of the sensor; you should see the value change instantly on the Serial monitor, as shown in the following screenshot:

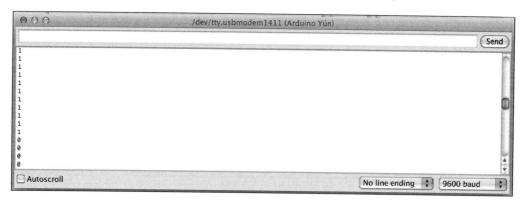

If you can see the values change instantly as you pass your hand in front of the sensor, this means that the Yún is wired correctly. You will also notice that the sensor turns red when it detects motion.

Now we are going to test the USB camera. We can actually test the camera without writing any Arduino sketch. What we are going to do instead is connect directly to the Yún board via SSH. Indeed, the camera is interfaced directly to the Linux machine of the Yún via the USB port, so the Arduino sketch will later have to use the `Bridge` library in order to access the camera.

For now, just go to a terminal window (the typical terminal that comes installed with OS X or Linux, or install one such as HyperTerminal if you are under Windows), and type the following command:

`ssh root@yourYunName.local`

Of course, you will have to put the name you gave to your own Yún in place of yourYunName. For example, the name of my Yún is myarduinoyun; therefore, I need to type myarduinoyun.local. This will establish a direct connection with the Linux machine of the Yún.

You will then be prompted to enter the password that you chose for your Yún. If it works, you should see the following screenshot being displayed on your terminal, which indicates that you are now working directly on the Yún:

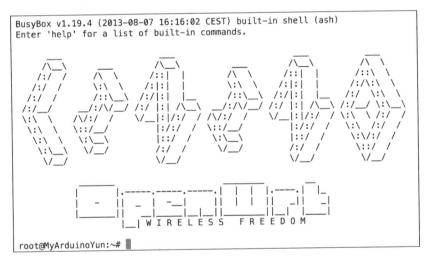

You can access all the functions from your Yún Linux machine. We are now going to install the required software for the camera. This requires the Arduino Yún to be connected to the Internet so that it can get the required packages, as described in the following steps:

1. The process starts by updating the package manager, opkg, as follows:

    ```
    opkg update
    ```

2. Install the UVC drivers, as follows:

    ```
    opkg install kmod-video-uvc
    ```

3. Install the python-openssl package that we will use later in the project, as shown in the following command:

    ```
    opkg install python-openssl
    ```

4. Finally, you can install the fswebcam software that we will use to take pictures, as shown in the following command:

    ```
    opkg install fswebcam
    ```

5. Once this part is done and the software is installed on the Yún, we can test the camera and take a picture. To also test whether the SD card is working at the same time, go over to the SD card folder, which is usually called sda1, using the following command:

    ```
    cd /mnt/sda1
    ```

6. You can now take a picture by typing the following command:

    ```
    fswebcam test.png
    ```

7. You should see some message being printed that starts with the following:

    ```
    --- Opening /dev/video0...
    Trying source module v4l2...
    /dev/video0 opened.
    ```

Some errors might be printed as well, but this doesn't matter for the process of taking a picture.

To check whether this works correctly, you can first check whether there is a file named test.png located on the SD card. To do this, you can simply type the following command:

```
ls
```

The preceding command will print the list of all the files in the current folder; in the present case, the SD card. You should see at least a file named test.png.

Now, to check that the picture is fine and not corrupted, you can, for example, remove the SD card from the Yún (by unmounting it first using the `unmount/dev/ sda1` command), and plug it directly to your computer using a micro SD card to normal SD card adapter. You should see the following screenshot in your browser (we already added the files that are required for the next sections of the project at this point, which explains the other files located on the SD card):

Name	▲	Date Modified	Size	Kind
▶ temboo		7 Mar 2014 10:21	--	Folder
test.png		Today 11:06	5 KB	PNG image
upload_picture.py		7 Mar 2014 10:47	970 bytes	Python Source

If you see a picture on your SD card at this point, open it to check that it was correctly taken. If that's the case, congratulations! Everything is now set up for you to write exciting applications with this project. If you can't see a picture at this point, the first step is to repeat the whole process again. Be careful to actually unmount the SD card after the picture is taken. You can also plug the camera directly to your computer to check whether the problem comes from the camera itself.

Recording pictures when motion is detected

The first application we are going to build with the hardware that we just step up will be only local, so nothing will be sent to the Web yet. In this section, we just want to build a camera that will be triggered by the motion sensor.

With this, you can, for example, check whether somebody entered your home while you were not there because the PIR motion sensor would instantly notice it. This section is really the foundation of the whole project. We are going to reuse the code developed in this section later when we write the piece of code to upload pictures to Dropbox.

For this part of the project, we don't want to use the SSH access to take pictures anymore; we need to trigger the camera right from the Arduino sketch. For this, we are going to use the `Bridge` library and the `Process` library to call a command on the Linux machine, just as if you were typing it on a terminal window.

The sketch starts by declaring the libraries that we need to use:

```
#include <Bridge.h>
#include <Process.h>
```

To call some commands on the Yún's Linux machine, we will need to declare a process, which is an object that we will call to emulate some terminal entries:

```
Process picture;
```

We'll also build a filename for each picture that will be taken, as shown in the following line of code. Indeed, we named the file `test.png` earlier, but in this application, we want every picture taken by the project to have a different name:

```
String filename;
```

Declare the pin on which the motion sensor is connected, as follows:

```
int pir_pin = 8;
```

We also need to define where the pictures will be stored. Remember, we want to store them all on the SD card, as follows:

```
String path = "/mnt/sda1/";
```

You can also store pictures locally on the Yún, but it would quickly saturate the memory of the Arduino Yún.

Then, in the `setup()` function, we start the bridge between the Atmel microcontroller and the Linux machine of the Yún, as follows:

```
Bridge.begin();
```

Also, we set the pin of the PIR motion sensor as an input, as follows:

```
pinMode(pir_pin, INPUT);
```

In the `loop()` function, what we want to do is to continuously read data from the motion sensor and trigger the camera if any motion is detected.

This is done by a simple `if` statement that checks the sensor's value, as follows:

```
if (digitalRead(pir_pin) == true)
```

Then, if some motion is detected, we need to prepare everything to take the picture. The first step is to build a filename that will contain the date on which the picture was taken. To do so, we are using the Linux date command that outputs the current date and time. This is important because we want to know what time the picture was taken at and give a unique filename to every picture. At the end, we also want to specify that this picture will be taken in a PNG format. The filename formatting part is done by the following code:

```
filename = "";
picture.runShellCommand("date +%s");
```

```
while(picture.running());

while (picture.available()>0) {
  char c = picture.read();
  filename += c;
  }
filename.trim();
filename += ".png";
```

Finally, we can take the picture. What we are going to do here is to call the `fswebcam` command again using the `runShellCommand` function of our picture process that will emulate a terminal entry.

We also want to use the maximum resolution available on the camera. In the case of the camera we chose, it was 1280 x 720 (standard HD resolution). We have quite a lot of space available on the SD card (4 GB with the one I used), so you can use the maximum resolution without running into problems. We recommend that you use a dedicated SD card for this project so that you don't run into problems with other files that could be stored on the card. For the sake of simplicity, we won't add an automated check to see whether the card is full, but this is something you should consider if you want to let the project run continuously over time. You can specify the resolution using the –o command at the end of the call. Finally, we can build the complete code to take a picture:

```
picture.runShellCommand("fswebcam " + path + filename + " -r
1280x720");
while(picture.running());
```

Note that we are also using a `while()` statement here to make sure that the `fswebcam` utility has enough time to take the picture. The complete code can be found at `https://github.com/openhomeautomation/geeky-projects-yun/tree/master/chapter3/triggered_camera`.

You can now upload the code to the Yún board and test the project. Once it's uploaded, try moving your hand in front of the sensor. The Arduino Yún should trigger the camera to take a picture and save it to the SD card. To make sure that a picture was taken at this point, you can simply check on the camera itself. For example, the Logitech webcam that I used has a small LED that turns green whenever it is active.

After a while, remove the SD card from the Arduino Yún (as earlier, unmount the SD card from the Yún first), and put it in your computer with the adapter we used earlier. You should see all the pictures that were taken at the root of the SD card, as shown in the following screenshot:

Name		Date Modified	Size	Kind
1395058857.png		Today 14:21	43 KB	PNG image
1395058862.png		Today 14:21	46 KB	PNG image
1395058869.png		Today 14:21	41 KB	PNG image
1395058875.png		Today 14:21	40 KB	PNG image
▶ temboo		7 Mar 2014 10:21	--	Folder
test.png		Today 11:06	5 KB	PNG image
upload_picture.py		7 Mar 2014 10:47	970 bytes	Python Source

Again, check these pictures to make sure that they are not corrupted and that everything worked as planned.

Sending pictures to Dropbox at regular intervals

We are now going to extend the code we built in the previous section and write some new code that automatically uploads the pictures that were taken by the camera to Dropbox. For this, we will need to build a slightly more complex software than in the previous part.

For now, we only used the Choreos (Temboo libraries) for the Arduino Yún. However, there are actually many other Choreos available for other languages, such as for Python. This is great news because the Linux machine of the Yún is capable of running Python code out of the box.

It's actually much easier to access the Dropbox API from Python, so that's what we are going to use in this part. We will build a Python script that uploads the pictures we took to Dropbox, and call this script from the Arduino sketch using the `Bridge` library and our picture processes.

I will now explain the content of the Python script. Later, we will save all these lines of code in a separate file, and put it on the SD card along with the Temboo Python SDK.

The Python script starts with the following lines of code:

```
from temboo.core.session import TembooSession
from temboo.Library.Dropbox.FilesAndMetadata import UploadFile
```

The Python script will also take an argument: the name of the file to be uploaded. This way, we can directly pass the name of file (built by the Arduino code with a timestamp) to the Python script. The following lines of code do exactly this:

```
with open(str(sys.argv[1]), "rb") as image_file:
    encoded_string = base64.b64encode(image_file.read())
```

Inside the script, you need to define your Temboo credentials, as follows:

```
session = TembooSession('yourTembooName', 'yourTembooApp',
'yourTembooKey')
```

These are exactly the same credentials we used for Temboo earlier. We then need to declare the upload file Choreo for Python that will be used to automatically upload pictures to Dropbox, as follows:

```
uploadFileChoreo = UploadFile(session)
uploadFileInputs = uploadFileChoreo.new_input_set()
```

The next step is to set the different inputs, which you had done when you created your Dropbox app, as follows:

```
uploadFileInputs.set_AppSecret("appSecret")
uploadFileInputs.set_AccessToken("accessToken")
uploadFileInputs.set_FileName(str(sys.argv[1]))
uploadFileInputs.set_AccessTokenSecret("accessTokenSecret")
uploadFileInputs.set_AppKey("appKey")
uploadFileInputs.set_FileContents(encoded_string)
uploadFileInputs.set_Root("sandbox")
```

Finally, we can order `uploadFileChoreo` to upload the file to your Dropbox folder in the corresponding folder of your app, as follows:

```
uploadFileResults =
uploadFileChoreo.execute_with_results(uploadFileInputs)
```

You can now save this code in a file named `upload_picture.py` and put it at the root of the SD card. Remember the Temboo Python library we downloaded earlier? It's time to unpack it and place it at the root of the SD card as well. Just make sure that it appears with the name `temboo` in the root of the SD card, so the Python file we just created can access it correctly. If no pictures have been recorded yet, the following screenshot shows what your SD card folder should look like:

Name		Date Modified	Size	Kind
▶ 📁 temboo	▲	7 Mar 2014 10:21	--	Folder
📄 upload_picture.py		Today 13:26	970 bytes	Python Source

We also need to slightly modify the Arduino sketch to upload pictures on Dropbox. We used exactly the same code base as in the previous section, so we will only detail the new code that was added.

In the part that is executed when motion is detected, right at the end of the loop, you need to use the picture process again to execute the Python script, as shown in the following code:

```
picture.runShellCommand("python " + path + "upload_picture.py " +
path + filename);
while(picture.running());
```

Note that we are passing along the same filename and path as the pictures that are recorded on the SD card, so the exact same picture name is recorded locally and sent to Dropbox.

The complete code for this part can be found at https://github.com/ openhomeautomation/geeky-projects-yun/tree/master/chapter3/dropbox_log.

You can now put the SD card back into the Arduino Yún, upload the updated Arduino sketch, and head to your Dropbox folder. At this point, you should note that a new folder was created in your Apps folder with the name of your Dropbox app that you set on the Dropbox website, as shown in the following screenshot:

Now, if motion is detected, the sketch should not only log the pictures on the SD card, but also on your Dropbox folder. If everything is working correctly, you can see that pictures arrive in real time inside your Dropbox folder as the Yún takes the pictures using the USB camera.

The cool aspect about these applications of our project is that this can be done from anywhere in the world. You can do this from your computer, of course, but also from a web browser. Many mobile devices can also run the mobile version of Dropbox, so you can see if somebody has entered your home right from your mobile device. On my computer, for example, Dropbox also sends me a notification that a new file was uploaded, so I can instantly see whether something is happening in my house and can act accordingly.

Live video streaming via Wi-Fi

To finish this chapter, we are going to see another cool application that we can make with the Arduino Yún and our USB camera. Remember that the camera is actually a standard webcam, and that it is also made to capture videos. Wouldn't it be cool to automatically stream video on a private video channel on the Web, so you can watch over your home in real time from anywhere just by going into a web browser? That's exactly what we are going to do in this section.

Many commercial IP cameras are actually doing this with proprietary solutions, but I wanted to use commonly available tools; this is why we chose the YouTube live event service to stream the video that can then be accessed from any device.

To make the application work, we first need to install some additional software packages on the Yún, as shown in the following steps:

1. Connect to the Yún again using SSH with your Arduino Yún name and password, and type the following command to get the correct package for live streaming:

   ```
   wget http://www.custommobileapps.com.au/downloads/mjpg-
   streamer.ipk
   ```

2. Note that if the link is not valid anymore and you can't find the files, this package is also available inside the code of this chapter. You can now install it with the following command:

   ```
   opkg install mjpg-streamer.ipk
   ```

3. You can now start the live streaming software on your Arduino Yún using the following command:

   ```
   mjpg_streamer -i "input_uvc.so -d /dev/video0 -r 640x480 -f
   25" -o "output_http.so -p 8080 -w /www/webcam" &
   ```

   Here, the parameter after -h is the resolution and the one after -i is the port on which the stream will be available. We also specified the number of frames per second using the -I command. The other options are less important and you do not have to worry about them.

Note that we didn't stream at HD resolution; it was apparently too much for the Arduino Yún, and the video stream suffered significant lag and also had corrupted images, which is not what we want at all. You can then access your stream by going to your Arduino Yún's address in your web browser followed by 8080 to specify the correct port. For example, in my case, it was http://myarduinoyun.local:8080/ stream.html.

This actually gives you direct access to the live stream. You should then see the stream interface with the live stream in the middle of the page, as shown in the following screenshot:

You can also use the different elements of the menu on the left to explore other possibilities of this streaming software. For example, you can get a link for **VideoLAN**, so you can access your stream right from the VLC player.

Now, this is already great and you could stop here to access your video stream from your local Wi-Fi network. But it would be even better if the stream was available online, so you could access it from anywhere in the world even without being connected to your local Wi-Fi network.

The first step is to go to your YouTube account in **VIDEO MANAGER** and to the **Live Events** menu on the left, as shown in the following screenshot:

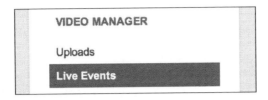

From this menu, you can create your stream just like you would create a new YouTube video. Make sure that you put the video as unlisted unless you want other people on YouTube to be able to see what's going on in your home. Compared to a private video, you will still be able to share the video with the people you know just by giving them the URL of the stream. Then, on the next page, YouTube will ask you to choose which encoder you want to use.

I chose **Wirecast** from the list and downloaded it from their website. In the Wirecast interface, you need to set the correct video source (by default, it will stream from your computer's webcam). In the menu where you can select the video source, select **Web Stream Source** and configure it, as shown in the following screenshot:

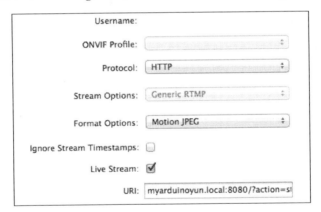

Basically, you need to choose HTTP as the protocol, use **Motion JPEG** as the format, and put the URL from the VideoLAN tab of the streaming interface. For example, for my project, it was `myarduinoyun.local:8080/?action=stream`.

After a moment, if everything is working fine, you should see the live stream from your USB camera appear directly in the main window of Wirecast. Don't worry if there is some lag at this point; it is only a delay usually; in my case, I had about 1-2 seconds of delay in the Wirecast software. The following is the image I got in the main Wirecast interface after adding the right video stream:

Also, make sure that this stream is the only one that will be sent to YouTube. For this purpose, delete all the other streams from the Wirecast interface. Indeed, by default, Wirecast puts the stream that comes from your webcam on the interface.

 Also note that using an HTTP stream is a feature from the paid version of Wirecast; it works perfectly in the free version, but you will get a watermark displayed on the video from time to time. Don't worry; it's actually not a problem to monitor what is going on in your house.

The next step is to actually stream data to YouTube. Click on the **Stream** button at the top of the interface, which should turn red, after which you will be prompted to enter your YouTube credentials. It should then automatically detect your live event video that you just created on YouTube.

Accept the settings, make sure it is streaming from Wirecast, and go back to the YouTube interface. You can now go to the video manager, and go to the **Live Control Room** tab. This is where you should see that YouTube is actually receiving some data from your Arduino Yún via Wirecast running on your computer. It should indicate that the **Stream Status** is **GOOD**, as shown in the following screenshot:

If this is not the case, please go back to the Wirecast application to check that the streaming process is working correctly. At this moment, don't worry; your stream is not working just yet. You should see that the **Preview** button, as shown in the following screenshot, is now available and can be clicked. Just click on it.

YouTube will then prepare your stream, as shown in the following screenshot, and you will have to wait for a moment (around 30 seconds when I tried it):

After a while, the page will be updated automatically so that you can move to the next step and actually start the streaming, as shown in the following screenshot:

Note that before making the stream live, you can preview it using the options on the preceding page. If what you see is satisfactory, you can now click on **Start Streaming** to finally finish the process. You will then have access to the stream on this page or directly on the dedicated page of the stream. The following screenshot is the final result on the YouTube interface:

You can see from the red dot below the video that the video is streaming live. Because the video is marked as **Unlisted**, only people with the URL can access it. You can, for example, mark the video as a favorite in your YouTube account and then access it from anywhere. You can also share it with your family and friends, so they can also watch the stream from their browsers.

Note that because we are using the Wirecast software on our computer to encode the stream for YouTube, we need to have our computer on for this to work. At the time this book was written, no software was available to directly stream the video to YouTube without the help of a computer, but this might change in the future, removing the need for a computer to stream the video.

Summary

Let's now summarize what we learned in this project. What we've built in this project is a security camera that can automatically log pictures locally and to Dropbox whenever motion is detected. We also learned how to stream the video that comes from this camera live on YouTube via Wi-Fi.

The following were the major takeaways from this project:

- In the first part of the project, we connected the USB camera to the Arduino Yún as well as the PIR motion sensor. We also plugged a micro SD card to the Yún so we can also record pictures locally, even if the Internet connection is not available. We also installed the required software packages for the project, such as the driver, to access the USB camera from a terminal command.

- Then, we tested our hardware by checking whether the motion sensor was working properly and by making sure that we can actually take pictures using the USB camera.

- Then, we built a simple application using our hardware to automatically take a picture when motion is detected and store it on the micro SD card. We tested this software on the Yún, and when we checked later, we found that some pictures were indeed stored on the SD card.

- After this simple application, we built on the sketch to automatically upload the pictures on our Dropbox folder as well. For this, we wrote a Python script that can automatically upload files to Dropbox. This script was then called from the Arduino sketch to upload pictures over to Dropbox whenever motion is detected.

- Finally, in the last part of the sketch, we used the video capabilities of the camera to stream video live on YouTube. This way, we can monitor what's going on in your home from wherever you are in the world; we just need an Internet connection and a device that can access YouTube, such as a smartphone.

Of course, there are many ways you can improve and extend this project based on what we learned in this chapter. You can, of course, have many of these modules with a camera and a motion sensor within your home. This way, you can have a complete security system that monitors your home remotely.

One way to improve the project is to integrate more Choreos into the project. For example, you can inject the Gmail Choreo we used in the first chapter to automatically send an e-mail alert as well whenever some motion is detected. This will create another layer of security in your home. In a similar way, you can also use the Twilio Choreo that will send you an SMS when motion is detected in your home.

You can also use the project for completely different purposes. While testing the project, we, for example, created a time-lapse device that takes pictures at regular time intervals (for example, every minute) and uploads these pictures on Dropbox.

In the next chapter, we are going to use the Arduino Yún's Wi-Fi capabilities for a completely different application: to control a mobile robot. We are going to use the power of the Arduino Yún to remotely control this robot and read out data that comes from the robot, all within a simple web application running on your computer.

4
Wi-Fi-controlled Mobile Robot

In this last chapter of the book, we are going to use the Arduino Yún in a completely different field: robotics. You will learn how to interface DC motors, as well as how to build your own mobile robot with the Arduino Yún as the brain of the robot, a distance sensor for the robot, and wireless control using Wi-Fi and a simple web interface. You will also be able to get a live display of the measurements done by the robot, for example, the distance that is measured in front of the robot by the ultrasonic sensor.

Building the mobile robot

Arduino boards are widely used in mobile robots because they are easy to interface with the different parts of a robot, such as sensors, actuators such as DC motors, and other components such as LCD screens. Arduino even released their own robot recently so people can experiment on a common robotic platform. These robots are usually programmed once and then left alone to perform certain tasks, such as moving around without hitting obstacles or picking up objects.

In this project, we are going to make things differently. What we want is to build a mobile robot that has the Arduino Yún as its "brain" and control it entirely via Wi-Fi from a computer or mobile device, such as a smartphone or a tablet. To do so, we will program an Arduino sketch for the robot that will receive commands and send data back, and program a graphical interface on your computer. This way, if you want to build more complex applications in the future, you simply need to change the software running on your computer and leave the robot untouched.

We are first going to build the robot using some basic mechanical and electrical parts. We will not only show you how to build the robot using a specific kit, but also give you a lot of advice on building your own robot using other equivalent components. To give you an idea about what we are going to build, the following is an image of the assembled robot:

At the bottom of the robot, you have most of the mechanical parts, such as the chassis, the wheels, the DC motors, and the ultrasonic sensor. You also have the battery at the center of the base of the robot. Then, you can see the different Arduino boards on top. Starting from the bottom, you have the Arduino Yún board, an Arduino Uno board, a motor shield, and a prototyping shield.

Assembling components in this project will be slightly different than before because we will actually have two Arduino boards in the project: the Yún, which will receive commands directly from the outside world, and an Arduino Uno board, which will be connected to the motor shield.

We will then perform the usual test on the individual parts of the robot, such as testing the two DC motors of the robot and the ultrasonic distance sensor that is located at the front of the robot. To test the motor, we are simply going to make them accelerate gradually to see whether or not the command circuit is working correctly. The measurements being received from the ultrasonic distance sensor will simply be displayed on the serial monitor.

The next step is to build the Arduino software that will receive commands from the computer and transmit them to the motors that move the robot around. At this point, we are also going to code the part that will transmit the distance information back to the computer. Because we want to standardize our code and make it usable by other projects, we will build this part with inspiration from the REST API of the Arduino Yún board that we already used in *Chapter 2, Creating a Remote Energy Monitoring and Control Device.*

Finally, we are going to build the server-side graphical interface on your computer, so you can easily control the robot from your computer or a mobile device and receive some data about the robot, such as the readings from the ultrasonic sensor. This server-side software will again use HTML to display the interface, JavaScript to handle the users' actions, and PHP to talk directly to your Arduino Yún board via the cURL function.

The required hardware and software components

You will need several mechanical and electrical components for this project apart from the Arduino Yún. The first set of components is for the robot itself. You basically need three things: a robot base or chassis that will support all the components, two DC motors with wheels so the robot can move around, and at least one ultrasonic sensor in front of the robot. We used a mobile robot kit from DFRobot (http://www.dfrobot.com/) that you can see in the following image:

The kit is called the **2 Wheels miniQ Balancing Robot chassis** and costs $32.20 at the time of writing this book. Of course, you don't need this kit specifically to build this project. As long as you have a kit that includes the three kinds of components we mentioned before, you are probably good to go on this project.

For the motors, note that the circuit we used in the motor shield can handle up to 12V DC, so use motors that are made to work at a voltage under 12V. Also, use motors that have an integrated speed reducer. This way, you will increase the available torque of your motors (to make the robot move more easily).

For the ultrasonic sensor, you have many options available. We used one that can be interfaced via the `pulseIn()` function of Arduino, so any sensor that works this way should be compatible with the code we will see in the rest of this chapter. The reference of this sensor at DFRobot is URM37. If you plan to use other kinds of distance sensors, such as sensors that work with the I2C interface, you will have to modify the code accordingly.

Then, you need an Arduino board that will directly interface with the DC motors via a motor shield. At this point, you might ask why we are not connecting all the components directly to the Arduino Yún without having another Arduino board in the middle. It is indeed possible to do with the sensors of the robot, but not the motors.

We can't connect the motors directly to an Arduino board; they usually require more current than what the Arduino pins can deliver. This is why we will use a motor shield that is specialized in that task. Usually, the Arduino Yún can't use these motor shields without being damaged, at least at the time of writing this book. This is due to the fact that motor shields are usually designed for Arduino Uno boards and the wrong pins on the shield can be connected to the wrong pins on the Yún. Of course, it would also be possible to do that with external components on a breadboard, but using a shield here really simplifies things.

This is why we will interface all the components with a standard Arduino board and then make the Yún board communicate with the standard Arduino board. We used a DFRduino board for this project, which is the name that DFRobot gave this clone of the Arduino Uno board. This is as shown in the following image:

Of course, any equivalent board will work as well, as long as it's compatible with the official Arduino Uno board. You could also use other boards, such as an Arduino Leonardo, but our code has not been tested on other boards.

Then, you need a motor shield to interface the two DC motors with the Arduino Uno board. We also used a motor shield from DFRobot for this project. The reference on the DFRobot website is **1A Motor Shield For Arduino**, as shown in the following image:

Again, most motor shields will work for this project. You basically need one shield that can command at least two motors. The shield also needs to be able to handle the motors you want to control in terms of voltage and current. In our case, we needed a shield that can handle the two 6V DC motors of the robot, with a maximum current of 1A.

Usually, you can look for motor shields that include the L293D motor driver IC. This integrated circuit is a chip dedicated to controlling DC motors. It can handle up to two 12V DC motors with 1A of current, which will work for the mobile robot we are trying to build here. Of course, if your shield can handle more current or voltage, that would work as well. The important point to look for is how to set the speed of the robot: the IC I mentioned can directly take a PWM command that comes from the Arduino board, so if you want to use the code prescribed in this chapter, you will need to use a shield that uses a similar type of command to set the motor's speed.

Finally, we added a simple prototyping shield on top of the robot to make power connections easier and so we can add more components in the future, as shown in the following image:

Again, you can use any equivalent prototyping shield, for example, the official prototype shield from Arduino. It is mainly so you don't have many cables lying around, but you can also use it to extend your robot project with more components, such as an accelerometer or a gyroscope.

You will also need a power source for your robot. As the DC motors can use quite a lot of current, we really recommend that you don't use power coming from your computer USB port when testing the robot or you will risk damaging it. That's why we will always use a battery when working with the motors of the robot. We used a 7.2V battery with a DC jack connector, so it can be easily inserted into the Arduino Uno board. This battery pack can also be found on the DFRobot website. You can also use some AA batteries instead of a battery pack. You will have to make sure that the total voltage of these batteries is greater than the nominal voltage of your DC motors.

As for the software itself, you don't need anything other than the Arduino IDE and a web server installed on your computer.

Robot assembly

It's now time to assemble the robot. We will show you the steps you need to follow on the robot kit we used for this project, but they can be applied to any other equivalent robot kit. The first step is to put the battery at the base of the robot, as shown in the following image:

Note that some metal spacers were also used at the base of the robot to maintain the battery in place and to provide support for the rest of the components. These spacers can also be found on the DFRobot website. Then, you can screw on two more spacers and the Arduino Yún board to the top of the chassis, as shown in the following image:

Then, we added the Arduino Uno compatible board on top of the two metallic spacers. At this point, you can screw on the Arduino Uno board; all the other components will just be plugged into these boards, as shown in the following image:

Then, you can simply plug the motor shield on top of the Arduino Uno board. At this point, you can also connect the cables that come from the DC motors to the motor shield screw headers. Be careful with this step; it is quite easy to plug the wrong cables from the DC motors. You need to connect each motor on a different connector on the motor shield board, as shown in the following image:

Finally, you can plug the prototyping shield on top of the robot. At this point, we already connected the ultrasonic sensor: ground goes to Arduino ground, VCC to Arduino's 5V pin on the prototype shield, and the signal pin goes into pin A0 of the Arduino board. If your ultrasonic sensor works with a digital interface, for example, you might want to use different pins. Please read the datasheet of your ultrasonic sensor for more information. The following image shows the state of the robot at this step:

Connecting the Arduino Yún and Uno boards

We are not done yet! For now, there are no connections between the Arduino Yún and the Arduino Uno board, so the Yún board won't be able to access the DC motors and the sensors of the robot. To solve this issue, the first step is to connect the power from the Arduino Uno board to the Yún board. This way, when we power the project using the battery, the Yún board will be powered as well.

To do so, simply connect the ground pins together and plug the V_{in} pin on the Arduino Yún to the 5V rail of the Arduino Uno, as shown in the following image:

To finish connecting the two Arduino boards, we need to connect them so they can speak together when the project is under operation. For this, we are going to use the I2C interface of the Arduino boards so they can send messages to each other. I2C stands for **Inter Integrated Circuit** and is a simple communication protocol that was developed for communication between circuits, and is widely used in electronics. There are two wires to connect for that purpose: SDA and SCL. To do so, simply connect pin 2 of the Yún board to pin A4 of the Uno board, and pin 3 of the Yún board to pin A5 of the Uno board, as shown in the following image:

The following image summarizes the connection between both boards:

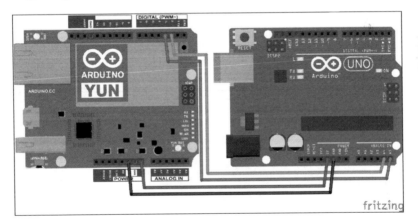

Finally, you can power up the project by inserting the DC jack connector of the battery into the power connector of the Uno board as shown in the following image:

If everything was done correctly in this step, you should see that both boards (the Yún and the Uno) are powered up, with some of their LEDs on. To help you build the robot, we also included two pictures of the sides of the robot that show you the different connections. The following is an image of a side of the robot that shows the power connections to the Yún:

The following image shows the connections from the I2C interface to the Yún:

Testing the robot's hardware connections

Before building the remote control part of the project, we want to make sure that the hardware is wired correctly, especially between the Arduino Uno board and the different motors and sensors. This is why we are first going to build a simple sketch for the Arduino Uno board to test the different components.

At this point, we are going to turn the motors of the robot on; so make sure the robot is standing on a small platform, for example, to prevent it from moving around while you are testing your different Arduino sketches with the USB cable connected to your computer.

The sketch starts by declaring the pins for the motors, as shown in the following code. Note that these pins are specifically for the motor shield we are using; please refer to the datasheet of your shield if you are using a different one.

```
int speed_motor1 = 6;
int speed_motor2 = 5;
int direction_motor1 = 7;
int direction_motor2 = 4;
```

Declare the pin used by the ultrasonic sensor as follows:

```
int distance_sensor = A0;
```

We also want to make the speed of the motor vary during operation, so we declare the variable as follows:

```
int robot_speed;
```

In the `setup()` part of the sketch, we need to specify that the motor pins will behave as output pins, as shown in the following code:

```
for(int i=4;i<=7;i++)
{
    pinMode(i, OUTPUT);
}
```

We also need to set a starting speed for the robot. Note that the speed of each motor will be set by PWM commands coming from the Arduino, so we have to specify a value between 0 (no voltage applied to the motor) and 255 (maximum voltage applied to the motor). Also, because of mechanical resistance on the motors, there is no linear relation between the value of the PWM command and the speed of the motor.

We used the value 75 as a starting speed, which is a very slow speed on our DC motors. However, depending on your own setup, this value will have a completely different effect. At this point, you can also experiment to see what the maximum PWM value is that will give you exactly zero speed on your DC motors. Make sure that the robot is not on the floor just yet as it would start to move forward and possibly damage things. We put it on a small stand so the wheels don't touch anything.

In the `loop()` part, everything is done by the function `send_motor_command`, which will be called for both motors. For example:

```
send_motor_command(speed_motor1,direction_motor1,robot_speed,1);
```

Let's see the details of this function. It starts by writing the speed of the motor on the correct pin as follows:

```
analogWrite(speed_pin,pwm);
```

Then, we need to set the the direction pin to the correct direction. This is done by a simple `digitalWrite` function as follows:

```
digitalWrite(direction_pin,dir);
```

Still in the `loop()` function, we call a function to measure the distance in front of the robot and print the result on the `Serial` port:

```
Serial.println(measure_distance(A0));
```

Let's see the details of this function. It starts by getting the raw measurement from the sensor using the `pulseIn` function. Basically, the sensor returns a pulse whose length is proportional to the measured distance. The length of the pulse is measured with the following function of Arduino dedicated for that purpose:

```
unsigned long DistanceMeasured=pulseIn(pin,LOW);
```

Then, we check whether the reading is valid and if it is, we convert it to centimeters using the following formula:

```
Distance=DistanceMeasured/50;
```

This is returned with the following code:

```
return Distance;
```

Finally, we update the speed at every iteration of the loop by increasing it by one unit, and we reset it if it reaches 255, as shown in the following code:

```
robot_speed++;
if (robot_speed > 255) {robot_speed = 75;}
```

The code for this section is available at the GitHub repository of the book and is stored in a file called `robot_test`: `https://github.com/openhomeautomation/geeky-projects-yun/tree/master/chapter4/robot_test`

It's now time to upload the code to the robot. Before doing so, please make sure that the robot is powered by the battery. Both motors of the robot should gradually accelerate upon reaching the maximum speed and then start again at a lower speed.

You can also open the serial monitor at this point to check the readings from the distance sensor. Try moving your hand or an object in front of the robot; you should see the distance changing accordingly on the serial monitor.

Building the Arduino sketch

It's now time to build the final sketch for our project. To be really precise, we should say sketches because we will have to develop two of them: one for the Uno board and one for the Yún board. You just have to make one simple change to the hardware at this point: connect the ultrasonic sensor directly to the Yún board by connecting the signal pin to the pin A0 of the Yún board.

Let's first focus on the Arduino Uno sketch. The sketch is inspired by the test sketch we wrote before, so it already includes the functions to control the two DC motors. To communicate between the two boards, we have to include the Wire library that is in charge of handling I2C communications:

```
#include <Wire.h>
```

Then, in the `setup()` part of the sketch, we need to declare that we are connecting to the I2C bus and start listening for incoming events. The Uno board will be configured as a slave, receiving commands from the Yún board, which will act as the master. This is done by the following piece of code:

```
Wire.begin(4);
Wire.onReceive(receiveEvent);
```

Let's see the details of this `receiveEvent` part, which is actually a function that is passed as an argument to the `onReceive()` function of the Wire library. This function will be called whenever an event is received on the I2C bus. What this function does is basically read the incoming data from the Yún, which has to follow a specific format like you can see in the following example:

```
speed_motor_1,direction_motor_1,speed_motor_2,direction_motor_2
```

For example, the first part of the previous message is read back with the following code:

```
int pwm1 = Wire.read();
Serial.print(pwm1);
char c = Wire.read();
Serial.print(c);
```

These commands that come from the Yún are then applied to the motors as follows:

```
send_motor_command(speed_motor1,direction_motor1,pwm1,dir1); send_
motor_command(speed_motor2,direction_motor2,pwm2,dir2);
```

Let's now focus on the Yún sketch. This sketch is inspired by the Bridge sketch that comes with the Arduino IDE and is based on the REST API of the Arduino Yún. To make things easier, we are going to create a new kind of REST call named robot. This way, we are going to be able to command the robot by executing calls like the following in your browser:

```
myarduinoyun.local/arduino/robot/stop
```

First, we need to include the correct libraries for the sketch as follows:

```
#include <Wire.h>
#include <Bridge.h>
#include <YunServer.h>
#include <YunClient.h>
```

Then, create a web server on the board:

```
YunServer server;
```

In the setup() function, we also join the I2C bus:

```
Wire.begin();
```

Then, we start the bridge:

```
Bridge.begin();
```

The setup() function ends by starting the web server as follows:

```
server.listenOnLocalhost();
server.begin();
```

Then, the loop() function consists of listening to incoming connections as follows:

```
YunClient client = server.accept();
```

The requests that come from these clients can be processed with the following command:

```
if (client) {
  // Process request
  process(client);

  // Close connection and free resources.
  client.stop();
}
```

If a client is connected, we process it to check whether or not a robot command was received, as follows:

```
String command = client.readStringUntil('/');

if (command == "robot") {
  robotCommand(client);
}
```

This function processes the REST call to see what we need to do with the motors of the robot. For example, let's consider the case where we want to make the robot go forward at full speed. We need to send the following message to the Arduino Uno board:

```
255,0,255,0
```

This is done by the following piece of code:

```
if (command == "fullfw") {
  Wire.beginTransmission(4);
  Wire.write(255);
  Wire.write(",");
  Wire.write(0);
  Wire.write(",");
  Wire.write(255);
  Wire.write(",");
  Wire.write(0);
  Wire.endTransmission();
}
```

We included three other commands for this simple REST API: stop (which obviously stops the robot), turnleft (which makes the robot turn left at moderate speed), turnright (which makes the robot turn right), and getdistance to return the distance coming from the ultrasonic sensor. We also inserted the measure_distance function in the sketch to read data that comes from the ultrasonic sensor.

We are now ready to upload the code to the robot. Remember that you have to upload two sketches here: one for the Uno board and one for the Yún board. The order doesn't matter that much, just upload the two Arduino sketches successfully by carefully ensuring that you are uploading the correct code to the correct board.

Both sketches are available in the following repository on GitHub: https://github.com/openhomeautomation/geeky-projects-yun/tree/master/chapter4/remote_control.

You can then test that the Yún board is correctly relaying commands to the Uno board. At this point, you can disconnect all cables and power the robot with the battery only. Then, go to a web browser and type the following code:

```
myarduinoyun.local/arduino/robot/turnright
```

The robot should instantly start turning to the right. To stop the robot, you can simply type the following code:

```
myarduinoyun.local/arduino/robot/stop
```

You can also type the following code:

```
myarduinoyun.local/arduino/robot/getdistance
```

This should print the value of the distance in front of the robot on your web browser. If you can see a realistic distance being printed on your web browser, it means that the command is working correctly.

Building the computer interface

We are now going to build an interface so you can control the robot remotely from your computer or a mobile device. This is actually quite similar to what we did for the relay control project, the main difference being that we also want to read some data back from the robot (in the present case the distance measurement from the ultrasonic sensor). There will be an HTML file that will host the different elements of the interface, some PHP code to communicate with the Yún board, some JavaScript to establish the link between HTML and PHP, and finally some CSS to give some style to the interface.

The first step is to create the HTML file that will be our access point to the robot control. This file basically hosts four buttons that we will use to control our robot and a field to continuously display the distance measured by the ultrasonic sensor. The buttons are declared inside a form; the following is the code for one button:

```
<input type="button" id="stop" class="commandButton" value="Stop"
onClick="stopRobot()"/>
```

The distance information will be displayed using the following line of code:

```
<div id="distanceDisplay"></div>
```

The following field will be updated with some JavaScript:

```
<script type="text/javascript">
  setInterval(function() {
    $("#distanceDisplay").load('get_distance.php');
  }, 1000);
</script>
```

Let's see the content of this PHP file. It basically makes a call to the REST API of the Yún board, and returns the answer to be displayed on the interface. Again, it will make use of the `curl` function of PHP.

It starts by making the cURL call to your Yún board with the `getdistance` parameter we defined in the sketch before:

```
$service_url = 'http://myarduinoyun.local/arduino/robot/getdistance';
```

It then prepares the call with the following code:

```
$curl = curl_init($service_url);
```

We get the answer with the following code:

```
$curl_response = curl_exec($curl);
```

We then print it back with the `echo` function of PHP:

```
echo $curl_response;
```

The PHP script that commands the motors is quite similar, so we won't detail it here.

Let's see the JavaScript file that handles the different buttons of the interface. Each button of the interface is basically linked to a JavaScript function that sends the correct parameter to the Arduino Yún, via the PHP file. For example, the `stop` button calls the following function:

```
function stopRobot(){
    $.get( "update_state.php", {command: "stop"} );
}
```

The same is done with the function to make the robot go full speed forward. To make it turn left or right, we can implement a more complex behavior. What we usually want is not for the robot to turn continuously by itself, but for example, to turn off a quarter of a turn. This is where the approach we took in this project becomes powerful. We can do that right on the server side without having to change the sketch on the Arduino board.

That's why to turn right for a given amount of time, for example, we will implement a series of commands on the server side and then stop. This is done by the following code:

```
function turnRight(){
    $.get( "update_state.php", { command: "turnright"} );
    sleep(350);
    $.get( "update_state.php", { command: "stop"} );
}
```

The `sleep` function itself is implemented in the same file and works by comparing the time that passed since the function was called, as shown in the following code:

```
function sleep(milliseconds) {
  var start = new Date().getTime();
  for (var i = 0; i < 1e7; i++) {
    if ((new Date().getTime() - start) > milliseconds){
      break;
    }
  }
}
```

Of course, we invite you to play with this sleep function to get the desired angle. For example, we set our sleep function such that the robot turns off about a quarter of a turn whenever we press the **Turn Right** button.

The code for the interface is available on the GitHub repository of the project: `https://github.com/openhomeautomation/geeky-projects-yun/tree/master/chapter4/remote_control`

Now, it's time to start the project. Be sure to place all the files at the root of your web server and make sure that the web server is running. Then, go to the folder of your web server in your browser (usually by typing `localhost`) and open the HTML file. The project also contains a CSS sheet to make the interface look better. The following is what you should see in your browser:

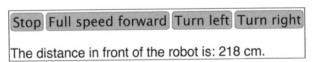

The field that displays the distance reading from the ultrasonic sensor should be updated automatically every second, so you can see whether or not this is working right away. Try moving your hand or an object in front of the robot and the value should change accordingly.

Before making the robot move around, we recommend that you test the different buttons while the robot is still on a small stand so it cannot move. Indeed, if something is wrongly coded on your server or within the Arduino sketch, your robot will not respond anymore and will randomly hit objects in your home.

You can now also test the different buttons. You can especially focus on the buttons that make the robot turn left or right and adjust the `sleep()` function in the PHP code to make them do exactly what you want. Notice that while your robot is moving around, the distance detected by the ultrasonic sensor in front of the robot is updated accordingly.

Summary

Let's see what the major takeaways of this chapter are:

- We started the project by building the robot from the different components, such as the robot base, the DC motors, the ultrasonic sensor, and the different Arduino boards.

- Then, we built a simple sketch to test the DC motors and the ultrasonic distance sensor.

- The next step was to build two Arduino sketches to control the robot remotely: one for the Arduino Uno board and one for the Yún board.

- At the end of the project, we built a simple web interface to control the robot remotely. The interface is composed of several buttons to make the robot move around, and one field that continuously displays the measurement that comes from the ultrasonic sensor mounted in front of the robot.

Let's now see what else you can do to improve this project. You can, for example, use the ultrasonic sensor data to make the robot act accordingly, for instance, to avoid hitting into walls.

Finally, you can also add many hardware components to the robot. The first thing you can do is add more ultrasonic sensors around the robot so you can detect obstacles to the sides of the robot as well. You can also imagine adding an accelerometer and/or a gyroscope to the robot so you will know exactly where it is going and at what speed.

You can even imagine combining the project with the one from the *Chapter 3, Making Your Own Cloud-connected Camera*, and plug a USB camera to the robot. This way, you can live stream what the robot is seeing while you control it with the web interface!

I hope this book gave you a good overview of what the Arduino Yún can add to your Arduino projects. Through the four projects in the book, we used the three main features of the Arduino Yún: the powerful embedded Linux machine, the onboard Wi-Fi connection, and the Temboo libraries to interface the board with web services. You can now use what you learned in this book to build your own applications based on the Arduino Yún!

Index

Thank you for buying
Internet of Things with the Arduino Yún

About Packt Publishing

Packt, pronounced 'packed', published its first book "*Mastering phpMyAdmin for Effective MySQL Management*" in April 2004 and subsequently continued to specialize in publishing highly focused books on specific technologies and solutions.

Our books and publications share the experiences of your fellow IT professionals in adapting and customizing today's systems, applications, and frameworks. Our solution based books give you the knowledge and power to customize the software and technologies you're using to get the job done. Packt books are more specific and less general than the IT books you have seen in the past. Our unique business model allows us to bring you more focused information, giving you more of what you need to know, and less of what you don't.

Packt is a modern, yet unique publishing company, which focuses on producing quality, cutting-edge books for communities of developers, administrators, and newbies alike. For more information, please visit our website: www.packtpub.com.

About Packt Open Source

In 2010, Packt launched two new brands, Packt Open Source and Packt Enterprise, in order to continue its focus on specialization. This book is part of the Packt Open Source brand, home to books published on software built around Open Source licenses, and offering information to anybody from advanced developers to budding web designers. The Open Source brand also runs Packt's Open Source Royalty Scheme, by which Packt gives a royalty to each Open Source project about whose software a book is sold.

Writing for Packt

We welcome all inquiries from people who are interested in authoring. Book proposals should be sent to author@packtpub.com. If your book idea is still at an early stage and you would like to discuss it first before writing a formal book proposal, contact us; one of our commissioning editors will get in touch with you.

We're not just looking for published authors; if you have strong technical skills but no writing experience, our experienced editors can help you develop a writing career, or simply get some additional reward for your expertise.

C Programming for Arduino

ISBN: 978-1-84951-758-4   Paperback: 512 pages

Learn how to program and use Arduino boards with a series of engaging examples, illustrating each core concept

1. Use Arduino boards in your own electronic hardware and software projects.

2. Sense the world by using several sensory components with your Arduino boards.

3. Create tangible and reactive interfaces with your computer.

4. Discover a world of creative wiring and coding fun!

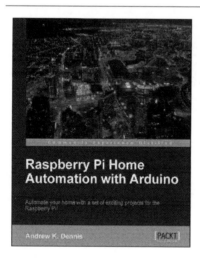

Raspberry Pi Home Automation with Arduino

ISBN: 978-1-84969-586-2   Paperback: 176 pages

Automate your home with a set of exciting projects for the Raspberry Pi!

1. Learn how to dynamically adjust your living environment with detailed step-by-step examples.

2. Discover how you can utilize the combined power of the Raspberry Pi and Arduino for your own projects.

3. Revolutionize the way you interact with your home on a daily basis.

Please check **www.PacktPub.com** for information on our titles

BeagleBone Robotic Projects

ISBN: 978-1-78355-932-9   Paperback: 244 pages

Create complex and exciting robotic projects with the BeagleBone Black

1. Get to grips with robotic systems.

2. Communicate with your robot and teach it to detect and respond to its environment.

3. Develop walking, rolling, swimming, and flying robots.

Learning ROS for Robotics Programming

ISBN: 978-1-78216-144-8   Paperback: 332 pages

A practical, instructive, and comprehensive guide to introduce yourself to ROS, the top-notch, leading robotics framework

1. Model your robot on a virtual world and learn how to simulate it.

2. Carry out state-of-the-art Computer Vision tasks.

3. Easy-to-follow, practical tutorials to program your own robots.

Please check **www.PacktPub.com** for information on our titles

48354772R00064

Made in the USA
Lexington, KY
26 December 2015